NEXT GENERATION
INNOVATION

KELLY CARNES

NEXT GENERATION
INNOVATION

SUPERCHARGE YOUR BUSINESS
THROUGH STRATEGIC
GOVERNMENT PARTNERSHIPS

Advantage | Books

Published by Advantage, Charleston, South Carolina.
Member of Advantage Media.

ADVANTAGE is a registered trademark, and the Advantage colophon is a trademark of Advantage Media Group, Inc.

Printed in the United States of America.

10 9 8 7 6 5 4 3 2 1

ISBN: 978-1-64225-807-3 (Paperback)
ISBN: 978-1-64225-806-6 (eBook)

LCCN: 2023910153

Cover design by Megan Elger.
Layout design by Wesley Strickland.

This publication is designed to provide accurate and authoritative information in regard to the subject matter covered. It is sold with the understanding that the publisher is not engaged in rendering legal, accounting, or other professional services. If legal advice or other expert assistance is required, the services of a competent professional person should be sought.

Advantage Books is an imprint of Advantage Media Group. Advantage Media helps busy entrepreneurs, CEOs, and leaders write and publish a book to grow their business and become the authority in their field. Advantage authors comprise an exclusive community of industry professionals, idea-makers, and thought leaders. For more information go to **advantagemedia.com**.

For President Bill Clinton and
Secretary Hillary Rodham Clinton,
whose pathbreaking 1992 campaign
first inspired me to enter public service.
And for my husband, Nathan,
who always has encouraged me
to take risks and try new things.

CONTENTS

Innovation—the advancement of technology and its application—is the engine of economic and productivity growth, a key to national security, and the driver of long-term improvements in living standards. And more than any country on Earth or in history, the United States has been the greatest driver and economic beneficiary of technology-based innovation, leveraged in a culture of entrepreneurship and resiliency.

—NATIONAL COMMISSION ON INNOVATION AND COMPETITIVENESS FRONTIERS

INTRODUCTION

AMERICA WAS FOUNDED on big ideas. Our country has long been home to risk-takers, disruptors, and innovators seeking to change the nation and the world. Over the years, our great national experiment has attracted big thinkers—and doers—from every walk of life. America would not be the advanced country it is today without these brilliant minds shaping our trajectory and bringing us a myriad of products, such as the internet, smartphones, and GPS technologies, and the tools used in our current fights against COVID and climate change.

The United States government has been there every step of the way with supportive policies and investments, helping big ideas become reality for over two hundred years. With a mission squarely aimed at prioritizing national security, health, energy, environmental protection, economic development, agriculture, and other societal needs, the government continues to play a critical role in funding and supporting technological progress in the public interest. United States government support has made possible many of the technologies we depend on and benefit from every day—including batteries for electric vehicles, bar codes, and technologies that increase our capacity to improve human life, for example, MRIs, genetic testing and tracing technologies, vaccines, and advanced prosthetics. From the perspective of an individual innovator, the United States government can provide grants ranging from a few hundred thousand dollars to multimillions to help accelerate technical progress and speed time to market through investments in research and development (R&D), infrastructure, and/or development of manufacturing capacity. In short, the United States government plays a catalytic role in bringing great innovations to the nation.

My passion for promoting American innovation developed decades ago when I served as an attorney representing technology companies in business transactions. In that role, I witnessed firsthand the birth of email services, the early stages of internet development, and other innovations in telecommunications. I supported a small company introducing one of the earliest voice recognition systems. So in a very small way, you may have me to thank for everyone's favorite invention: the modern call center, in which you talk to a computer rather than a person when you have an issue or complaint.

1925	The civilian aviation industry
1945	The Doppler radar
	The flu shot
1946	MRIs
1953	Supercomputers
1958	Microchips
1962	LED lights
1966	Autonomous robots
1967	3D seismic imaging
1971	Closed captioning
1974	Barcodes
1976	Computer simulation software
1977	Modern hydraulic fracturing
1980	Modern wind energy
1985	Enriched infant formula
1986	The hepatitis B vaccine
1988	The Human Genome Project
1990	Genetic tracing services
1992	Smartphones
1993	GPS
1995	The internet
1998	Google
	Touch screens
2004	The kidney matching program
	Self-driving cars
2006	The HPV vaccine
	Modern upper-limb prosthetics
2007	Modern lower-limb prosthetics
2008	Weather apps
2010	Tesla cars
2011	Automated Personal Assistants
2017	Super-absorbent oil sponge
2020	The Moderna coronavirus vaccine

Later, I served as the Senate-confirmed Assistant Secretary of Commerce for Technology Policy. Our mission at the Office of Technology Policy was to promote the competitiveness of United States industry, which included supporting wise investments in R&D, talent, and infrastructure. In this position, I was privileged to work alongside many great public servants and role models. Our job in the Office of Technology Policy was to communicate with the technology business community and to develop, advocate for, and implement programs and policies to promote innovation and American competitiveness, including increased investment in research & development in R&D, developing and nurturing the nation's pool of science, technology, engineering, and mathematics (STEM) talent, supporting development and expansion of technology-related infrastructure (including our stellar network of national laboratories and shared user facilities), and promoting a pro-innovation business climate.

One important program at that time was the Advanced Technology Program (ATP), adopted under the George H. W. Bush Administration, expanded by President Clinton, and managed by the National Institute for Standards and Technology (NIST). The ATP made investments in individual companies and consortia—frequently in partnership with laboratories and/or universities—to advance high-risk enabling technologies with the potential for extraordinary economic benefits to the United States. At the time, this program was controversial, because large companies could benefit from the funding even though those companies were most likely to be able to fund projects themselves. However, the results were impressive; one review conducted in 2001 found that 25 percent of projects resulted in expanded production capacity, 41 percent completed production prototypes for at least one application, 35 percent completed pilot production or a commercial demonstration, and 12 percent moved into actual production. At that time, 10 percent of companies had earned early revenues, and ATP awardees had filed 105 patent applications and 7 copyrights and had been issued 11 patents.[1]

I recall that one of my mentors, Dr. Mary Good, who served as Under Secretary of Technology at the Commerce Department and who was a real trailblazer—she earned a Ph.D. in chemistry, served as the head of a multibillion-dollar R&D organization in private industry, and then served as a longtime government advisor at a time when women were not prominent in these positions—spent a substantial amount of time with Members of Congress answering questions about this controversial program such as "Why is the government providing IBM with funding? They can easily afford to make these investments without United States government help."

Inevitably, Dr. Good would say, in a very feisty way, something to the effect of "I want IBM to make the investments here in America. Don't you?"

That principle of encouraging investments in America is just as valid today. Unfortunately, after quite some time, ATP was discontinued due to lack of political support. That story is worth noting for the lesson it suggests. When Bill Clinton became President in the early 1990s, he made the ATP one of his signature programs, encouraging its growth into over a billion-dollar-a-year enterprise. Republicans in Congress took a firm stance against it in their attempts to reduce or eliminate programs that President Clinton favored. Although those efforts did weaken the program over the years, they didn't eliminate it. It wasn't until 2006 that ATP was discontinued. The lesson I believe it's important to draw from the politicization of the ATP is *not* that the unpredictability associated with political maneuverings stands as

---

1 Glenn R. Fong, 2001, "Repositioning the Advanced Technology Program," Issues in Science & Technology 18 (1): 65. https://search.ebscohost.com/login.aspx?direct=true&AuthType=ip,shib&db=a9h&AN=5477 282&site=ehost-live&scope=site.

an argument against pursuing government support. Instead, I believe the important lesson this suggests is this: if a program exists and is funded today, entrepreneurs should not hesitate to utilize its resources!

Another major program to which I had the opportunity to contribute was the Partnership for a New Generation of Vehicles, launched in 1993 (PNGV). This was a collaboration among the (then) Big Three automakers (General Motors, Ford, and Daimler Chrysler) and a whopping thirteen government agencies and departments seeking to develop technologies needed to make a midsize family sedan three times more fuel efficient. The relevant technologies included advancements in internal combustion engines, designing and creating electric vehicles and their batteries, development of new lighter-weight materials, and exploration of alternative power sources such as hydrogen and flywheels. PNGV was a seminal program that led to the deployment of new generations of vehicles by the automotive industry and launched trends in automotive development and manufacturing that continued from that point forward and have grown even stronger today.

The last story I will share for now from my time in government service is the thrill of participating in the launch of the National Nanotechnology Initiative (NNI) in fiscal year (FY) 2001.[2] The NNI—which is still going strong today—focuses on coordination of nanotechnology R&D efforts across eleven government agencies. In the FY 2023 budget, the NNI was funded at over \$2 billion. The NNI also has created several very valuable user facilities open to the nanotechnology research community. While not all United States progress in nanotechnology over the past twenty-two years can be attributed to the NNI, this initiative has certainly had a major impact. Some

---

2   A nanometer is one-billionth of a meter. For reference, a sheet of paper is about a hundred thousand nanometers thick.

of the developments in the industry include core processor technology, clean technologies, and everyday products such as sunscreen, adhesives, and heat and water-resistant fibers. Various nano-based products have been commercialized, and the industry stands on the precipice of even greater contributions across many fields, including transportation, energy, and medicine, to name just a few.

During my years in law and government, I learned a great deal about the challenges facing entrepreneurs (whether in start-ups, venture-backed companies, or innovators working on leading-edge technologies within large global companies) when it comes to moving a concept from inception to commercialization. I also learned that partnerships among industry, academia, government, and nonprofit organizations can play a critical role in overcoming those challenges. In government, I also learned quite a bit about political maneuverings and how even the best-laid plans to bring new innovations to market can sometimes be derailed if inadequate attention is paid to building support across the large universe of players who influence United States government technology programs and policies.

When Administrations changed and I left government service, I remained keenly aware that many organizations of all sizes and capabilities were not leveraging United States government policies, programs, and funding to their fullest potential. That was when I decided to launch TechVision21 to help a wide range of clients advance exciting technologies to increase their success while also helping improve American competitiveness, national security, job creation, and economic growth. This "doing well while doing good" philosophy underpins all of TechVision21's work. From the start, I sought out—and was quickly found by—enterprises that were highly motivated to promote the national good and had made that motivation part and parcel of their overall mission. At TechVision21, we have been very fortunate to work directly with both the Federal and state governments, with companies of all sizes and stages of development, and with universities and nonprofit organizations to advance their technologies and policy interests.

One of the things I truly love about my work is the opportunity to collaborate with an expert team that is deeply knowledgeable about theories and models of innovation, the history of science and technology policy, and details of United States government programs and the budget-making process. This array of expertise gives our team the collective insight needed quickly to pinpoint each client's specific needs and understand how to address those needs through innovative partnerships with government and policy-shaping initiatives. In over twenty years in business, my team and I have supported numerous pathbreaking technologies, including information and communications technology, defense, broadband, a wide variety of renewable energy technologies, semiconductors, automotive technologies, nanotechnology, artificial intelligence, and more.

As my team and I look to the future, we are excited about the opportunities available to Americans to ride the next wave of innovation. In my opinion, there are four powerful forces at work that make the need for next-generation innovation even more urgent and imperative when it comes to enhancing Americans' quality of life and providing our children and grandchildren with extraordinary opportunities to achieve the American Dream.

SHARE OF UNITED STATES HOUSEHOLDS USING SPECIFIC TECHNOLOGIES

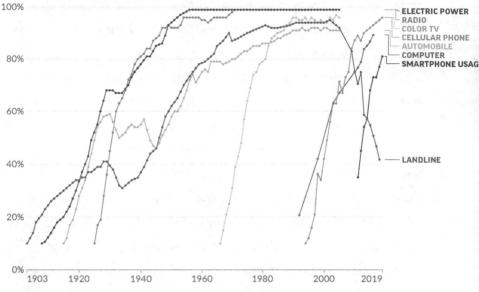

Source: Horace Dediu; Comin and Hobijn (2004); other sources collated by Our World in Data   OurWorldInData.org/technological-change • CC BY

1. The pace of technological change and technology adoption has accelerated dramatically over the past few decades.

2. Many technologies that have been under development for years or decades are poised for imminent commercialization and impact. These include quantum computing, artificial intelligence, biotechnology, nanotechnology, next-generation microelectronics, and advanced manufacturing. Today, we are looking at a world that is set to change dramatically as these exciting technologies are brought to market. Artificial intelligence alone is estimated to contribute almost $16 trillion to the global economy by 2030.[3]

3. Globalization and international competition for R&D capabilities and talent are advancing at a breakneck pace. Whereas the

---

3   PricewaterhouseCoopers, "PWC's Global Artificial Intelligence Study: Sizing the Prize," PwC, accessed March 13, 2023, https://www.pwc.com/gx/en/issues/data-and-analytics/publications/artificial-intelligence-study.html.

United States once performed over 60 percent of the world's R&D, several countries have dramatically increased their R&D investment, and many countries today have adopted technology programs and policies to improve the well-being and quality of life of their citizens. Many of these countries have patterned their efforts after successful United States initiatives.

For many years, large numbers of foreign-born students educated in science and engineering at American universities remained in the United States, to the great benefit of our nation. Today, many of these students return to their native countries, often in response to aggressive recruitment incentives, applying what they have learned to strengthen their country's capabilities in innovation and technology development. China stands out among nations with its announced intention to become the world's preeminent technological and military power and its commitment to historic levels of investment to achieve these goals.

The Made in China 2025 initiative, announced in 2015, seeks to transform China from a manufacturing giant into a global science and technology power by 2049 (the 100th anniversary of the People's Republic of China), while it set a target to become one of the most innovative countries by 2020 and a leading innovator by 2030. Made in China targets advanced IT, advanced machine tools, robotics, aerospace technology, maritime equipment, new energy vehicles, biomedicine, and advanced medical equipment.

China is targeting development of the entire semiconductor ecosystem, including spending more than $150 billion over 10 years for investments and acquisitions ... China's semiconductor policies include a strong government role in developing international rules for protection of intellectual property, advancing Chinese standards, use of antitrust authorities and priority financing vehicles ... The Chinese government uses production targets; subsidies; tax preferences;

trade and investment barriers (including pressure to engage in joint ventures); and discriminatory antitrust, IP, procurement and standards practices.[4]

—Council on Competitiveness Commission on Innovation and Competitiveness Frontiers, Competing in the Next Economy, 2020

4. Fourth, and most important, the challenges the United States and the world face today—including addressing climate change, mitigating the effects of increasingly severe weather events (such as tornadoes, hurricanes, severe wildfires, and floods), controlling (or hopefully avoiding) the next pandemic, and ensuring adequate supplies of fresh water and food for the world's growing population—are daunting and demand big "Manhattan Project" type solutions.

The significance of the Manhattan Project, the controversial R&D undertaking that produced the first nuclear weapons, is that it involved an enormous collaborative effort among the Federal government, industry, and the scientific sector. It is estimated that more than six hundred thousand people worked on the project.[5]

Recognition of the scale and scope of these four challenges and opportunities is starting to have a dramatic influence on the government policy and program arena. When I was in government in the 1990s, we made it a priority to educate decision-makers about the power of technology to improve people's lives. Our team published numerous papers and reports pointing out that leading economists had

---

4   "Competing in the Next Economy," Compete, March 1, 2022, 17. https://compete.org/2020/12/14/competing-in-the-next-economy/.

5   "The Manhattan Project," Nuclear Museum, accessed March 13, 2023. https://ahf.nuclearmuseum.org/ahf/history/manhattan-project/.

long credited advances in technology with creating over 50 percent of the nation's long-term economic growth. Despite that evidence, and the obvious nature of the critical role government can and does play in advancing innovation, we found there was a lot of time spent in Washington debating what the government's proper role should be in the technology policy arena. At its core, the debate centered around the Federal government's role in influencing or controlling the nation's economy (with technology as an important driving force). At times, it seemed that the debate over whether government's role should expand or decrease took on near-religious fervor.

I am happy to say that decision-makers have been far more pragmatic and action-oriented in recent years; there is currently bipartisan support for investing much more aggressively in R&D, building the nation's STEM talent, and creating programs, tax incentives, and other mechanisms that make it more likely that leading-edge technology will be developed and deployed in the United States. In fact, Congress and the Biden Administration took many great ideas that have been discussed and debated over the past few decades and put these ideas into several recent pieces of legislation. As a result, there has been an expansion of Federal investment in areas such as electric vehicles and EV charging, grid-scale and vehicle energy storage, energy-efficient buildings, renewable energy, upgrades to the national electric grid, climate change mitigation, carbon capture, broadband, biotech and the bioeconomy, cybersecurity, and semiconductors.

There also has been a significant expansion in types of government engagement, migrating at least somewhat from a more traditional role investing primarily in early-stage R&D (leaving to the private sector the rest of the work of moving innovations to market) to include additional support for closer-to-market activities such as pilot and demonstration projects, and scale up to overcome what's known as the *valley of death*—that point in the development cycle where funding that is somewhat more plentiful in early stages often dries up—and for commercialization and advanced manufacturing.

LEGISLATION ENACTED IN RECENT YEARS SETS THE STAGE FOR LARGE AND STRATEGIC GOVERNMENT INVESTMENTS IN LEADING-EDGE TECHNOLOGY. EXAMPLES INCLUDE:

The **Bipartisan Infrastructure Law (BIL)**, enacted in November 2021, includes more than a half trillion in new technology-related spending over the next few years, including $326 billion for roads and bridges; $82 billion for public transportation; $25 billion for airports; $16 billion for ports and waterways; $38 billion for transportation safety; $18 billion for electric vehicles, buses, and ferries; $64 billion for water projects; $38 billion for resilience; $75 billion in clean energy and power; and $64 billion for broadband. Goals of BIL include reducing the United States' emissions by 50 percent or greater from 2005 levels by 2030, creating a 100 percent carbon pollution-free power sector by 2035, and achieving a net-zero economy by 2050.

The **Inflation Reduction Act of 2022** invests over $360 billion in energy security and climate change programs over ten years. Goals of the legislation include lowering energy costs, increasing clean energy production, and reducing carbon emissions by about 40 percent by 2030.

The **Creating Helpful Incentives to Produce Semiconductors (CHIPS) and Science Act**, enacted in August 2022, appropriated over $50 billion to incentivize investment in facilities and equipment in the United States for semiconductor fabrication, assembly, testing, advanced packaging, R&D, and workforce development. In addition, the act authorized the establishment of regional innovation and technology hubs as well as new and expanded investments in STEM education and training from K–12 to community college, undergraduate, and graduate education.

The **Fiscal 2023 Appropriations Bill**, enacted in December 2022, provided a historic level of funding—$9.9 billion—for the National Science Foundation to support additional grant-making capability and up to thirty-five thousand more scientists, technicians, teachers, and students.

This convergence of urgent economic, national security, and societal needs; the relentless pace of new technology development and adoption; today's global competitive environment; and policymakers' willingness to enable government to take a more active role in helping move forward the latest innovations creates what I believe is a golden age of opportunity for American innovators to partner with government. The amount of current and planned investment in science and technology over the next five to ten years is staggering. Planned investments in "green energy" alone are estimated to reach up to $3 trillion in the next decade.[6]

I have two major goals in writing this book. The first is to persuade current and future generations of American innovators to consider partnerships with government as a strategic option for advancing your own commercial interests while also benefiting the country and world. The second is to offer practical and common-sense suggestions, based on lessons learned from my own and my team's experience about what to do (and what to avoid).

In a nutshell, these include the following:

- Performing in-depth research to determine whether working with the government is right for your organization and to understand how your organization's goals are best aligned with government needs and interests.

---

6   Jennifer Hiller, "Clean Energy Push Promotes Backlash," The Wall Street Journal (New York, New York) May 8, 2023.

- Once your organization makes a decision to seek government funding or partnerships, do your homework, perform in-depth opportunities research, set priorities, and develop a strategy and compelling messaging.

- Meet and build relationships with program managers and policymakers whose work is aligned with your interests.

- Participate in the competitive process.

I want us to ask ourselves every day, how are we using technology to make a real difference in people's lives.

—PRESIDENT BARACK OBAMA

CHAPTER 1

IS WORKING WITH THE FEDERAL GOVERNMENT RIGHT FOR YOU?

FOR MANY INNOVATORS, SEEKING GOVERNMENT funding and partnerships can seem like a daunting challenge.

Where do I start? Who are the right people to talk to in government? What types of inventions will various government departments find appealing? Why would the government choose to partner with my organization?

I recognize that some people feel reluctant to partner with the United States government, due, at least in part, to the relatively longer timeframes and higher levels of bureaucracy required to bring new technologies to fruition when compared with the private sector. Some may perceive partnering with foreign governments to be easier. Almost every entrepreneurial company with which I have worked has been approached by foreign interests, and in some cases, the foreign governments were ready to write a check on the spot. I certainly have witnessed companies with innovative products and services look toward China, at the very least when it comes time for manufacturing. This, of course, brings its own set of challenges.

Another major deterrent faced by many American researchers and technology companies is lack of awareness of the multiple ways they might partner with government agencies to accelerate the development and deployment of innovative technologies. Over the years, I have seen that many high-tech start-ups and small business innovators have promising ideas, but they struggle to gain the capital sufficient to advance and demonstrate their innovations. Too often, these technologies fall into what, as I mentioned in the Introduction, is called the *valley of death*—they never reach the commercialization process, and the innovative companies that developed them are stymied along the path to growth and success.

As someone who is eager to encourage fruitful partnerships between innovative companies and the Federal government, I want to walk you through a series of questions that can help determine whether such a partnership is right for your project or business overall. Anyone seriously considering working with the Federal government should answer some big-picture questions, the first of which is this: *Does seeking a government partnership make sense for my organization?*

An honest answer to this question will involve some self-reflection, taking into consideration organizational goals, projects, and current (or intended) products, as well as existing and planned capabilities. Well before looking to see which agencies or organizations might be compatible with your project, determining whether you should work with government at all involves an assessment of your own capabilities, mission, and motives. It is one thing to want to advance your technology and accelerate the commercialization process. It is something more to believe that your project or your business proposal will aid the United States in matters of national security or global competitiveness or support other goals such as health, space, addressing climate change, or similar goals.

Very generally, if your business need matches a national need or addresses a national concern, you should consider working with the United States government. When making this determination, keep in mind that one ongoing and overarching national need is to ensure that the work of developing, commercializing, and manufacturing beneficial projects and products happens *within* the United States. There are many good reasons to pursue foreign partnerships, and some agencies gladly support these efforts, but there must always be a broad consideration of how American interests are addressed. American innovators and firms also need to weigh the anticipated benefits of foreign partnerships against detractors such as human rights or health and safety concerns.

It is one thing to want to advance your technology and accelerate the commercialization process. It is something more to believe that your project or your business proposal will aid the United States in matters of national security or global competitiveness or support other goals such as health, space, addressing climate change, or similar goals.

If and when you determine that, at least in this very general sense, you are likely to be a good match for the Federal government, your next question should be something like this: *Is my project or business prepared to work with the Federal government?*

There are several relevant considerations here:

HOW WELL DOES YOUR TECHNOLOGY OR PRODUCT MEET GOVERNMENT-WIDE AND AGENCY-SPECIFIC MISSION NEEDS?

Federal R&D programs are, for the most part, mission driven, so your project needs to address clearly the stated goals of any particular program.

IS YOUR WORK OR YOUR PRODUCT STATE OF THE ART IN YOUR FIELD?

If your answer is no, you may not be a clear partner for the government. It is important to offer a strong value proposition by clearly showing that whatever project you are bringing to the table is either unique or unusual, has measurable value—whether that is meeting an agency mission goal, speeding time to market, reducing costs, simplifying manufacturing processes, creating a whole new product or service that does not yet exist, or otherwise helping advance the field in some meaningful and measurable way that has the potential for positively impacting society and the economy—*and* is better than most, if not all, of your competitors. You may need to update your market research to answer this question thoroughly.

WHAT IS YOUR ORGANIZATION'S CURRENT STATUS IN THE TECHNOLOGY DEVELOPMENT CYCLE?

Would participating in a government program help you advance from where you are now to having a product that is ready for commercialization to meet a market window? Precisely what kind of progress might you be able to make? For example, if you are a very small company, government funds might allow you to hire another couple of people with specialized expertise to help accelerate your progress.

Alternatively, if you are part of an orphaned project within a large company, unsure about how to move your project forward or get the attention it needs to move forward, government funding or partnership might help accelerate the project from a technical standpoint or may otherwise shine a light on the excellent work you are doing.

HOW FLEXIBLE CAN YOU BE IN MEETING YOUR ORGANIZATION'S GOALS?

If you are in a position to work with various kinds of partners—if you are a small company, these could be academic institutions, large companies that could potentially help commercialize your product, or nonprofit organizations that might be able to tap into economic-development or manufacturing-related support—then you can expand the array of Federal opportunities that might work to advance your technology.

DO YOU HAVE THE STAYING POWER NECESSARY FOR SUCCESS?

Although it does happen, it is unusual for an organization to strike gold on its first try working with government. To ensure success, those seeking partnerships need to designate the work of establishing those partnerships as a priority within their organizations, designate at least one person who can make this his or her focus—at least half time—for a year to eighteen months, and ensure there are adequate personnel and financial resources backing that person up.

Finally, it is important to realize that, as with other matters in life, there are no guarantees. Unfortunately, it is possible to do everything right and still not gain traction. I could point to numerous cases of innovators who came to me with game-changing products and services that should have succeeded and delivered dramatic impact, yet they faced challenges gaining momentum with potential Federal government partners. In cases where innovators fail, there are many reasons, but two resonate with me: bias built into the system and disconnect on policy priorities. Technology almost always moves faster than laws, regulations, and sometimes human understanding.

The lesson here is that when an innovative organization brings a fresh approach to the table, that organization may encounter bias or other limitations on understanding. In my experience, the thornier the problem and more innovative the approach, the more resistance your organization may face. It might be that government representatives have already seen prior generations of this proposed technology and make certain assumptions about how this new concept is much like the old one. Or they might completely misunderstand the marketplace or value of a proposed technology.

BRINGING CHANGE TO THE AUTOMOTIVE INDUSTRY

Traditionally, one of the industries most resistant to change has been the United States automotive industry. A major government initiative focused its efforts on overcoming that resistance and addressing significant technical challenges faced by the industry.

Our team worked with several companies that offered innovative solutions in the automotive industry, including companies developing novel internal combustion engines and nanotechnology-based batteries. They were led by extremely capable, experienced, and visionary executives. These executives had managed to solve problems that were considered intractable by incumbent industry players such as the Detroit Three companies as well as by program managers at the Department of Energy. These companies had their acts together but still struggled to secure funding even though their efforts to educate their audiences were excellent and thorough. The main reason for this outcome was that policies in place during the Obama Administration preferred vehicle electrification and that key Federal employees held preconceived notions about what certain types of internal combustion engines and certain battery chemistries could and could not do. Key players inside and outside of government did not understand the full value these innovative companies had to offer and could not see clearly the problems they would solve. It is important to note that while these companies did not succeed in attracting funding, they did change the conversation. Today, several years later, the Departments of Energy and Defense have indicated an openness to the same types of novel engines and batteries proposed by these extraordinary innovators. They were truly ahead of their time.

In these cases, and when a product or idea is truly revolutionary, it will be important to include education as a key element of your strategy for success with the Federal government. Educating the audience for your ideas and products enables you to utilize the views of policymakers inside and outside of government—such as think tanks—to help explain your value proposition and amplify your message. We will talk more about this strategy in a later chapter, but for now I will share a story of a long-term educational effort that eventually paid off.

One of my favorite projects over the years was the opportunity to support Ocean Thermal Energy Technology (OTEC). OTEC uses the temperature differential between warm ocean surface water and the cold deep ocean to generate electricity, hydrogen, and/or fresh water. The technology has enormous potential to generate low-cost baseload electricity; this is extremely important for contributing to the electric grid, as most renewable technologies provide only intermittent power (for solar, while the sun is shining, and for wind, while the wind blows). The proponent of the technology was a large global company with significant access to resources. But the technology was expensive to build.

In the late 1970s, American researchers were able to demonstrate the first OTEC plant on a Navy barge off the coast of Hawaii, and then in the late 1990s, establish a larger plant in Hawaii. In the early 2000s, because of significant advances in ocean engineering during the intervening years, proponents of the technology believed that OTEC could become economically viable, with important applications for islands and other locations far from the traditional electric grid.[7] Proponents of the technology engaged in a massive educational effort, not only with relevant Federal agencies and people on Capitol Hill but even within the company itself. With the help of my team, they succeeded in convincing the Department of Energy that OTEC could be a viable energy solution and even won funding for R&D on critical components. Although still not a mainstream technology for energy generation, work on OTEC continues, with ongoing efforts to develop a platform that can withstand extreme weather and overcome other key economic barriers to large-scale deployment.

---

7   In addition, much thought has been given to how OTEC might someday support the American electric grid. The Carter Administration invested a few hundred million dollars in OTEC and surveyed the Gulf of Mexico, identifying numerous sites where OTEC might be appropriately deployed in the Gulf.

OCEAN THERMAL ENERGY CONVERSION OTEC TECHNOLOGY

GENERATOR

WORKING FLUID

TURBINE

PLATFORM

OCEAN SURFACE

POWER
CABLE

HEAT
EXCHANGE
CONDENSER

PUMP

WARM WATER INTAKE
25°C (77°F)

PUMP

HEAT
EXCHANGE
EVAPORATOR

SEPERATE WARM
WATER DISCHARGE
23°C (73°F)

SEPERATE COLD
WATER DISCHARGE
7.5°C (46°F)

OR COMBINED DISCHARGE
16°C (61°F)

COLD WATER INTAKE
5°C (41°F)

Source: National Oceanic and Atmospheric Administration (NOAA). "Ocean Thermal Energy Conversion (OTEC) Technology." Accessed March 13, 2023. https://coast.noaa.gov/data/czm/media/technicalfactsheet.pdf.

In the coming chapters, we will not leave behind the questions posed here regarding whether partnering with government could be a good fit for your organization or project. Instead, the strategy suggestions for successful interaction with government that follow should help you continue to clarify both whether, and the extent to which, a Federal government partnership is right for you.

The need to educate potential partners can seem both time consuming and daunting, especially when you are coming to the task already deeply immersed in and committed to the innovative technology that you are developing. My recommendation to you is this: always look for and take advantage of opportunities to educate. And following that, don't let day-to-day politics keep you from pursuing what you want.

One of the most successful companies I have ever seen has always engaged with potential government partners and engaged on Capitol Hill, come rain or shine.

Year after year, no matter who is in power, no matter what the funding levels and specific programs are at a given moment, this company goes out and pursues its interests across a variety of areas. There are years when it doesn't have any successes and years when it has big successes. In part, its efforts are not limited to keeping its bottom line growing. This company knows that there are other reasons for engaging that are equally important as securing funding: influencing the development of a new program within an agency, influencing changes within the policy environment, or contributing to shifting the conversation around a given technology and its benefits to the nation. Educating potential partners is, in this sense and nearly always, a good in itself.

TAKE ACTION

Identify your preliminary answers to the following four questions:

1. How well does your technology or product meeting agency or government-wide mission needs?

2. Is your work or your product state of the art in your field?

3. What is your organization's current status in the R&D cycle?

4. How flexible can you be in meeting your organization's goals?

5. Do you have the staying power necessary for success?

Together, let us explore the stars, conquer the deserts, eradicate disease, tap the ocean depths, and encourage the arts and commerce.

—JOHN F. KENNEDY

CHAPTER 2

THE FEDERAL ROLE IN SCIENCE AND TECHNOLOGY– A BRIEF HISTORY

WHEN I WAS IN GOVERNMENT, I FOUND it impressive when organization's representatives reveal in their conversations with government agency representatives that they understand the historical trajectory in which the technology they are developing is embedded.

I believe there is a business case to be made for having at least some knowledge of the history of Federal engagement in technological advances. It is in that vein that I offer you the following summary.

I also believe there is an advantage to recognizing patterns and government funding and policy priorities. Knowing the lineage of partnerships related to the technology on which you work can help you understand the Federal government's current policies as well as where these might be headed in the future. For these reasons, I believe there is a business case for having at least some knowledge of the history of Federal engagement in technological advances. It is in that vein that I offer you the following brief summary.

The Federal government has supported R&D since the founding of our nation,[8] although initially at a small scale, and focused on specific matters of direct interest to the government, such as land surveys and exploration.[9] In the mid-1800s and following the Civil War, the Federal government took on new responsibilities, including performing agricultural and meteorological R&D, and efforts to promote food health and safety. From there, you can almost see how government prioritization of different issues occurs in response to the need to find technical solutions to pressing national problems. The creation of the National Bureau of Standards (now the National Institute of Standards and Technology) occurred in 1901 to create and maintain standards of weights and measures. That was at least partly in response to some very practical problems, including the inability to couple railroad cars together because they were all made to different standards and the need to connect firehoses. The creation of standard measures allowed these problems to be solved.

Health research was a major focus for quite some time, leading to the creation of the National Institutes of Health (NIH) in 1931. The second New Deal, beginning in 1935, fostered a major expansion of Federal R&D programs as the government sought to use technology to address major social and economic challenges.

Science and technology played a game-changing role in World War II, most notably the Manhattan Project, which developed the first nuclear weapons. This was followed by postwar increases in

---

8   America's commitment to innovation is written into the Constitution. Article I, Section 8, Clause 8 provides that [The Congress shall have power] "To promote the progress of science and useful arts, by securing for limited times to authors and inventors the exclusive right to their respective writings and discoveries."

9   Richard Rowberg, "Federal R&D Funding: A Concise History," CRS Report for Congress. Congressional Research Service, August 14, 1998, https://www.everycrsreport.com/files/19980814_95-1209_5099a810 54a63d58f79d6d18b4572fe7270f5a2e.pdf.

R&D to improve the nation's defense—including the creation of the National Military Establishment in 1947, retitled the Department of Defense (DOD) in 1949—as well as health, well-being, and economic security.

By the end of World War II, the United States government was a powerhouse supporter of innovation, investing in over 60 percent of the world's R&D. In 1950, the United States created the National Science Foundation to "promote the progress of science; to advance the national health, prosperity, and welfare; to secure the national defense."[10] This facilitated a new era of government-sponsored R&D, contributing greatly to our nation's defense, health, economy, and quality of life.

Americans did not just generate ideas; they put them into action.

For many years following, this strategy worked to make America the most innovative nation in the world. Our country has been a world leader because we coupled government and private investment in R&D with the power of United States industry to commercialize new technology. In other words, Americans did not just generate ideas; they put them into action. Other countries, weakened by war, could not match the capabilities of the United States.

In the 1960s, research priorities shifted somewhat with the start of the space program—a response to the Soviet Union's launch of Sputnik in 1957. The National Aeronautics and Space Administration (NASA) was established to put a man on the moon by the end of the 1960s. By the end of that decade, NASA had become the second-largest R&D funding agency after DOD.[11]

---

10   "About NSF," NSF, accessed March 13, 2023, https://www.nsf.gov/about/.

11   Rowberg, "Federal R&D Funding: A Concise History, 1998."

In the 1970s, R&D funding growth resulted from Federal policy actions in three areas:

- The oil embargoes of 1973 and 1979 stimulated a large increase in energy R&D, leading to what eventually became the Department of Energy.

- The launching of the war on cancer in 1971 resulted in a large increase in R&D at the NIH for the first half of the decade.

- Increased concern about the Soviet Union led to a major buildup in defense capabilities and rapid growth of R&D spending by the DOD.

In the 1980s, the military buildup continued with even greater emphasis on technological superiority, rapidly increasing DOD R&D. At the same time, the Reagan Administration's desire to reduce Federal involvement in the economy reduced support for developing products with commercial potential.[12]

As a result of these two policy actions, funding for defense R&D constituted 61 percent of all Federal R&D by 1988 compared to 45 percent in 1980.[13] In the mid-1980s, the appearance of AIDS also resulted in a rapid increase in health R&D. Nearly all this work was supported by NIH as Federal public health policy assigned a major role to research as part of its response to the disease. AIDS funding accounted for 18.5 percent of the overall 28.8 percent increase in NIH funding.[14]

---

12   "A Brief History of the Department of Energy," Energy.gov, accessed March 13, 2023. https://www.energy.gov/lm/brief-history-department-energy#:~:text=The%20Energy%20Crisis%20and%20the%20Department%20of%20Energy&text=In%201977%2C%20the%20establishment%20of,and%20balanced%20national%20energy%20plan.

13   Rowberg, "Federal R&D Funding: A Concise History, 1998."

14   "4 Supporting the NIH AIDS Research Program—NCBI BOOKSHELF," accessed March 13, 2023. https://www.ncbi.nlm.nih.gov/books/NBK234085/.

In the late part of the 1980s, through the start of the Clinton Administration, the Federal government once again shifted policy on funding R&D that directly supported development of new or improved technologies and processes with potential commercial applications. A significant expansion of funding took place for joint government–private-sector projects. The principal agencies involved in these efforts were NIST, the DOE, and the DOD.

Within DOE, programs in applied energy technology development began to receive more money, and greater emphasis was placed on technology transfer activities, particularly at the national laboratories. A principal tool of the latter was the Cooperative Research and Development Agreement (CRADA), which allowed the private sector and Federal government researchers to join together on a project of potential benefit for the private partner(s). In DOD, dual-use technology development—technologies with both civilian and defense applications—was emphasized, both to expand the commercial manufacturing base for producing military products and to help exploit military technology for civilian purposes.

In the time since then, the landscape shifted. Numerous other countries can and do compete with the United States in both R&D and manufacturing. Moreover, other nations—such as China—are beginning to rival the United States in making R&D, business operations and manufacturing affordable. The result is that many previously United States–based companies have taken their production processes overseas to nations that offer lower costs and stronger economic incentives.

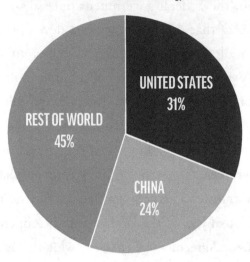

GROSS DOMESTIC EXPENDITURE ON R&D
Source: OECD Main Science and Technology Indicators

UNITED STATES 31%

REST OF WORLD 45%

CHINA 24%

In recent decades, there has been a growing awareness that America cannot dominate global science and technology as it once did. Today, all countries have access to new knowledge and emerging technologies. As other nations have increased their R&D investments, the U.S. global share has dropped to 31 percent (2019), and China alone now accounts for 24 percent of global R&D spending.[15] American R&D investment as a percentage of GDP ranks fourth behind Israel, South Korea, and Taiwan (2021).[16].

That said, the past five Presidents—Clinton, Bush, Obama, Trump, and Biden—all have recognized and supported the development and commercialization of new technologies, ranging from health, next-generation vehicles to wind and solar energy, biotechnology, quantum computing, artificial intelligence, autonomous systems, semiconductors, space technologies and systems, and strengthening manufacturing capabilities in the United States.

15   OECD Main Science and Technology Indicators (extracted May 11, 2022).

16   OECD Main Science and Technology Indicators. In addition, at least one recent study has concluded that the rate of technical progress in key sectors (including semiconductors and pharmaceuticals) may be slowing, based on a review of patent data. While the number of patents has increased exponentially, fewer "disruptive" innovations are emerging. https://www.nature.com/articles/s41586-022-05543-x

President Clinton: Bill Clinton and running mate Al Gore were the first candidates in history to release a platform including positions on innovation policy. Clinton and Gore promised to harness the power of technology to improve the quality of American life and reinvigorate the economy. Among its many accomplishments, the Clinton Administration increased funding for civilian R&D by 43 percent; made significant investments in R&D—across numerous disciplines, including nanotechnology and biomedical research—created a twenty-first century research fund, extended the Research and Experimentation Tax Credit; dramatically increased funding for the NSF and NIH; provided significant support for the Human Genome Initiative; and presided over the first comprehensive telecommunications reform legislation in sixty years, which lowered prices, increased customer choice, and sped the deployment of new technology. President Clinton also launched the Partnership for a New Generation of Vehicles with a goal to triple the fuel efficiency of then-current vehicles, and the National Nanotechnology Initiative. President Clinton and Vice President Gore also made promoting the internet a top priority. They launched the highly effective "Internet 2" program, which connected 100+ universities to the internet at speeds up to 1000 times higher than the then current internet. Perhaps most important, in 1997, President Clinton released a strategy for promoting global e-commerce, notable in its philosophy of allowing e-commerce to grow as much as possible without regulation or government interference.

I ask you simply to imagine that new century, full of its promise, molded by science, shaped by technology, powered by knowledge. These potent transforming forces can give us lives fuller and richer than we have ever known. They can be used for good or ill.

If we are to make the most of this new century, we—all of us, each and every one of us, regardless of our background—must work to master these forces with vision and wisdom and determination. The past half-century has seen mankind split the atom, splice genes, create the microchip, explore the heavens. We enter the next century propelled by new and stunning developments.

Just in the past year we saw the cloning of Dolly the sheep, the Hubble telescope bringing into focus dark corners of the cosmos never seen before, innovations in computer technology and communications, creating what Bill Gates calls the world's new 'digital nervous system,' and now cures for our most dreaded diseases— diabetes, cystic fibrosis, repair for spinal cord injuries. These miracles actually seem within reach.

—PRESIDENT BILL CLINTON[17]

President George W. Bush: President Bush continued the Clinton Administration's efforts on vehicle fuel economy and made major investments in R&D funding to bring hydrogen and fuel cell technology from the laboratory to the automobile showroom. President Bush also made major strides in improving health information technology and electronic records. He believed that innovations in electronic health records and the secure exchange of medical information would help transform healthcare in America by improving healthcare quality, preventing medical errors, reducing healthcare costs, improving administrative efficiencies, reducing paperwork, and increasing access to affordable care. The Bush Administration also made investments in R&D and the deployment of broadband technology a priority, including budgeting over $2 billion for the National Information Technology Research and Development program to support United States science, engineering, and technology leadership and to bolster American economic competitiveness.

This country needs a national goal for … the spread of broadband technology. We ought to have … universal, affordable access for broadband technology by the year 2007, and then we ought to make sure as soon as possible thereafter, consumers have got plenty of choices when it comes to [their] broadband carrier.

—PRESIDENT GEORGE W. BUSH[18]

---

17   "Commencement Address at Morgan State University in Baltimore, Maryland May 18, 1997," Public Papers of the Presidents of the United States, GovInfo, accessed March 13, 2023. https://www.govinfo.gov/content/pkg/PPP-1997-book1/pdf/PPP-1997-book1.pdf.

18   National Archives and Records Administration, accessed March 13, 2023. https://georgewbush-white-house.archives.gov/infocus/technology/economic_policy200404/chap4.html.

President Obama: President Obama's goals included upgrading government use of technology, net neutrality, and stimulus spending on technology, electronic medical records, and advanced manufacturing. The Administration's signature American Recovery and Reinvestment Act (ARRA) appropriated approximately $100 billion to renewable energy projects (including solar energy and advanced batteries), broadband development and deployment, and next-generation manufacturing. One notable program under the Obama Administration is the Network of Manufacturing USA Institutes, which brings together industry, academia, and Federal partners within advanced manufacturing institutes to increase American manufacturing competitiveness and promote a robust and sustainable national manufacturing R&D infrastructure.

> We have to do everything we can to encourage the entrepreneurial spirit, wherever we find it. We should be helping American companies compete and sell their products all over the world. We should be making it easier and faster to turn new ideas into new jobs and new businesses. And we should knock down any barriers that stand in the way. Because if we're going to create jobs now and in the future, we're going to have to out-build and out-educate and out-innovate every other country on Earth.

> —PRESIDENT BARACK OBAMA, SEPTEMBER 16, 2011[19]

President Trump: President Trump continued the trend of increasing civilian R&D investment, focusing on, among other issues, reducing the cost of broadband deployment, improving digital health information, integrating drones into the American air traffic system, and promoting the safe deployment of autonomous vehicles into the United States transportation system. A signature achievement of the Trump presidency—supporting the development of vaccines against the COVID-19 virus in less than one year from the identification of the virus—is unprecedented in the history of vaccinology. This

---

19   "Technology," National Archives and Records Administration, National Archives and Records Administration, accessed March 13, 2023. https://obamawhitehouse.archives.gov/issues/technology.

achievement also demonstrates the power of long-term investment in R&D; pharmaceutical companies were able to commercialize and deploy life-saving vaccines by building on decades of prior research in mRNA technologies.

We're on the verge of new technological revolutions that could improve, virtually, every aspect of our lives, create vast new wealth for American workers and families, and open up bold, new frontiers in science, medicine, and communication.

–PRESIDENT DONALD J. TRUMP[20]

President Biden: President Biden has advocated for, and overseen, a dramatic increase in funding for R&D, energy and climate programs, broadband deployment, semiconductor-related R&D, manufacturing, facilities creation and expansion, infrastructure, and workforce development, and programs to combat climate change. As previously noted in the Introduction, several major pieces of pro-innovation legislation have recently been enacted on President Biden's watch. These include the Bipartisan Infrastructure Law of 2021, the Inflation Reduction Act of 2022, the CHIPS and Science Act of 2022 and the Fiscal Year 2023 Appropriations Bill. This substantial commitment of the President and Congress to supporting American innovation makes today a great—and unprecedented, at least in recent years—time for American innovators to work with government. President Biden has set very aggressive goals for deployment of renewal energy resources. For example, he has called for operation of the United States electric grid solely on clean energy by 2035.[21]

I will make no apologies that we are investing to make America strong. Investing in American innovation, in industries that will define the future, and that China's government is intent on dom-

---

20   National Archives and Records Administration, remarks by President Trump at American Leadership in Emerging Technology Event. June 2017, accessed March 13, 2023. https://trumpwhitehouse.archives. gov/briefings-statements/remarks-President-trump-american-leadership-emerging-technology-event/.

21   Dino Grandoni, "Biden's renewable energy goals blow up against a painful WWII legacy," *The Washington Post*, May 3, 2023.

inating. Investing in our alliances and working with our allies to protect our advanced technologies so they're not used against us. Modernizing our military to safeguard stability and deter aggression. Today, we're in the strongest position in decades to compete with China or anyone else in the world.

–PRESIDENT JOSEPH R. BIDEN[22]

As we look to the future, the Federal government is faced with an imperative: to maintain and expand the United States' position as a global leader in innovation and technological advancement. In my opinion, this presents not only a new set of opportunities but a new urgency to make good on the nation's long-standing reputation for innovation and its entrepreneurial spirit. America remains in a position of strength given its investments in research infrastructure, high-tech manufacturing, and the basic research that seeds future innovations.

In this critical moment in our nation's history, will your ideas and projects become the ones that help reposition the country's ability to control its economic destiny and set the standard for global competition?

If the history of Federal engagement in science and technology has one thing to teach us, it is that the path toward the practical application of ideas and research is often indirect and shaped by larger debates focused on societal and political concerns. With a view of how the environment around innovators shapes their work, their discoveries, and the pace with which those discoveries come to market, it may be easier not only to see how your work fits into the scheme of our nation's past but to gain some perspective on the real-world applications of your research and knowledge and the value of persistence on behalf of nation- and world-transforming technology.

In a 2023 address at Georgetown University's School of Foreign Service, Secretary of Commerce Gina Raimondo offered the following assessment of the possibilities presented by the 2022 CHIPS and Science Act:

Let's think about what's possible 10 years from now if we are bold. We can show the world that efficient global supply chains do not require us to sacrifice resiliency and security. We can once again lead in manufacturing, and all of the innovation that grows from it. The level of technological leadership, supplier diversity, and resiliency we are seeking does not and will not exist anywhere else in the world. It will create the new generation of innovators who will write the next chapter in our history.[23]

TAKE ACTION

Dig deeper into the historical trajectory of Federal involvement in your area of expertise, listing up to three insights that you can draw upon as you pursue Federal partnerships.

---

23   "Remarks by U.S. Secretary of Commerce Gina Raimondo: The Chips Act and a Long-Term Vision for America's Technological Leadership," delivered at Georgetown University's School of Foreign Service on February 23, 2023, US Department of Commerce, March 3, 2023. https://www.commerce.gov/news/speeches/2023/02/remarks-us-secretary-commerce-gina-raimondo-chips-act-and-long-term-vision.

As commercial and cross-sector innovation gained pace, governments have gone beyond fixing market failures. In addition to helping strengthen strategic sectors such as defense and space, governments are fostering cross-sector solutions for a myriad of societal challenges, including public health, climate change, and cybersecurity.

—DELOITTE, "GOVERNMENT AS CATALYST"[24]

CHAPTER 3

UNDERSTANDING TODAY'S FEDERAL SCIENCE AND TECHNOLOGY ENTERPRISE

IT IS POSSIBLE for nearly any project to have a component that fits within the purview of the Department of Defense. That is both because the DOD is the largest government agency in the United States and because the DOD is responsible for more than the technologies that move us forward in terms of defending ourselves and establishing American military leadership. The DOD and the Services also work in communities all over the country and the world helping service members and their families live their lives. The DOD, due to its responsibility for addressing local challenges for its members, has a large R&D footprint.[24]

---

24 "Government Transformation," Deloitte Insights, Deloitte, March 24, 2022. https://www.deloitte.com/global/en/our-thinking/insights/industry/government-public-services/government-trends/2022/government-transformation.html.

When you are considering how to optimize and maximize your opportunities (an approach I highly recommend you take), you will want to understand how your work can apply across multiple agencies.

If you are primarily interested in agriculture, you will certainly look at partnerships available to you within the Department of Agriculture. But then you will want to see if there are defense-related applications for your work. When you are considering how to optimize and maximize your opportunities (an approach I highly recommend you take), you will want to figure out how your work can apply across multiple agencies. Of course, determining your ability to work across the board will involve reflecting on the resources, time, and priorities of your organization; its mission; and its goals. In an ideal situation, you would be able to pursue several opportunities at once, addressing all issues where your work might have an impact, raising a discussion point, or offering a novel approach. If you are resource constrained, as most of us are, you will start with the obvious agencies and programs. Even if that is your approach, it is still good to know early on how you might extend your reach and participation from there.

One of the great joys during the eight years I worked at the Department of Commerce was working across Federal R&D agencies on common challenges and initiatives. The Federal government's capabilities are formidable, and when leveraged properly, can deliver very impressive results. This is one thing that President Clinton and Vice President Gore did extremely well. It always seemed to me that establishing collaborative enterprises was part of their DNA. As mentioned earlier, they were the youngest Presidential and Vice-Presidential candidates, and they saw technology as an important pathway into the future. The pair of them campaigned on big initiatives, such as PNGV and ATP, and they put emphasis on the importance of investing in R&D, as well as commercialization and the pathways that lead to the economic and societal impact of technology.

One of the projects with which I was routinely engaged, and from which I drew a considerable amount of satisfaction, was a joint effort among the Departments of Commerce, Education, and Energy and the National Science Foundation to ensure an adequate supply of highly qualified practitioners in Science, Technology, Engineering and Math (STEM) fields at all levels. One major way to achieve this goal was to encourage more women and minorities to choose STEM degrees and professions. When it came to encouraging women (and this is still true today), the task was to address issues of choice: whereas women graduated high school adequately as well prepared as men to pursue STEM majors in college for a complex host of reasons, they did not choose these fields and so lagged behind their male counterparts.

When it came to encouraging minorities and people of color, the issue was a matter of preparation: many of these students were not provided as many opportunities to prepare for STEM careers in high school and therefore frequently needed both more preparation and encouragement in order to enter the STEM pipeline. Once prepared to pursue college-level STEM courses, minority students and students of color chose—and continue to choose—STEM fields as often as white men. Besides creating internship and one-on-one mentoring programs, the coalition of government departments also put great effort into offering students alternative images of scientists (to counter the paradigm of a middle-aged white man wearing a white lab coat holding a beaker of some fluorescent substance) and engaging students in a wide variety of enjoyable activities. Promoting STEM careers remains a passion of mine. Our team has been very fortunate to work on several projects to help the U.S. military to assess their own needs for STEM talent and help develop pathways to create a pipeline at all skill levels from which these military services can benefit.

One exciting program completely run by the private sector, that has shown outstanding results over the last few decades is For Inspiration and Recognition of Science and Technology (FIRST). Founded by the extraordinary innovator Dean Kamen (who holds more than 100 patents and is the inventor of the Segway, stair climbing wheel chairs, and an insulin pump for diabetics), FIRST brings the joy of sports to robotics. Teams use standard parts kits and their own imagination to build robots that can play a variety of games. Competitions are held annually, and winners are crowned in several categories. FIRST has a strong track record of students pursuing engineering in college, and many scholarships are available. I served as a FIRST judge for many years, and I credit this program, at least in part, for sparking my own children's success in science and technology.[25]

As you launch your plans to work in partnership with government, it is important to have and leverage a general understanding of the overall enterprise. It is rare that any one agency or program will offer everything your organization needs. Furthermore, as you meet with agency representatives, you will want to demonstrate an understanding of how the overall enterprise functions and where your projects or products could contribute to meeting national needs.

Today's Federal Science and Technology Enterprise spans numerous departments, agencies, and programs, and is focused on maintaining or establishing American leadership in critical areas of technology. In what follows, I offer you a brief description of the roles of some key agencies with which you may want to become familiar.

---

25   https://www.firstinspires.org/

DEFENSE DEPARTMENT

The Department of Defense (DOD), including the military services—the Army, Air Force, Navy, and now the Space Force—has robust short-, medium-, and long-term R&D and commercialization agendas focused on ensuring that the American military remains the world's best equipped and most agile and effective military force.

Because the DOD and its services have a large footprint, both in the United States and abroad, these agencies also are concerned with a wide variety of issues one would not normally consider defense-related, such as healthcare, housing, energy use, and a wide variety of issues contributing to the safety and quality of life of service members.

DOD entities are estimated to receive $92.5 billion in R&D funding for FY 2023.

HEALTH AND HUMAN SERVICES DEPARTMENT

What is today the Department of Health and Human Services (HHS) has developed over the years a very sophisticated program of R&D to promote drug discovery, drug development, and commercialization and to make other investments to support human health and well-being. For fiscal year 2023, the National Institutes of Health (NIH) is allocated over $46 billion for R&D to advance science and speed the development of new therapies, diagnostics, and preventive measures. Among other priorities, this includes funding for Alzheimer's disease, support for research on opioids and pain, women's health, autoimmune disease, and addressing health disparities.

DEPARTMENT OF ENERGY

The mission of the Department of Energy (DOE) is to ensure America's security and prosperity by addressing its energy, environmental, and nuclear challenges through transformative science and technology solutions. Given increasing concerns regarding climate change, the DOE's renewable energy portfolio has grown dramatically over the past few decades. DOE has received almost $22 billion in R&D appropriations for FY 2023. Areas of focus include programs in solar, wind, vehicles, and energy efficiency, as well as support for more traditional energy sources.

DEPARTMENT OF TRANSPORTATION

DOT R&D programs are focused on promoting safety and mobility and creating and enhancing infrastructure and environmental protection. The R&D budget was about $200 million in 2020.

DEPARTMENT OF AGRICULTURE

Among its many missions, the USDA performs R&D dedicated to the creation of a safe, sustainable, competitive American food system, as well as strong communities, families, and youth through integrated research, analysis, and education. The agency's R&D budget was over $3 billion in FY 2022.

ENVIRONMENTAL PROTECTION AGENCY

The agenda of the EPA is quite broad, spanning efforts to address a wide variety of environmental challenges, encompassing efforts on air and water pollution, land waste and cleanup, pesticides, chemicals, and lead remediation, to name a few. The EPA has a regulatory role (creating rules to require a variety of organizations to take or abstain from actions in support of a clean and healthy environment). In addition, it has a strong role in supporting science and technology. The agency's proposed R&D budget for 2023 was $11.88 billion.

DEPARTMENT OF COMMERCE

Through its 13 bureaus, the Department of Commerce works to drive U.S. economic competitiveness, strengthen domestic industry, and spur job growth in U.S. communities. It includes the National Institute of Standards and Technology which conducts a wide range of research on emerging technologies such as quantum computing, artificial intelligence, advanced materials, bioscience, advanced electronics, and advanced manufacturing. NIST oversees the Hollings Manufacturing Extension Partnership, Manufacturing USA Institutes, and implementation of the CHIPS Act to strengthen U.S. semiconductor manufacturing. The National Oceanic and Atmospheric Administration advances commercial space and conducts R&D on climate, oceans, and marine fisheries. The Economic Development Administration provides grants to stimulate technology-based economic growth, and the National Telecommunications and Information Administration plays a key role in U.S. broadband deployment. The department includes the Patent and Trademark Office, plays a key role in international trade and export controls, produces economic statistics, and carries out the U.S. census. In Fiscal Year 2022, the DOC's R&D budget was estimated to exceed $2 billion.

NATIONAL AERONAUTICS AND SPACE ADMINISTRATION

NASA carries out the U.S. civil space program and American space exploration. At 20 centers and facilities across the United States, NASA conducts research on Earth's climate, the sun, and solar system. It also performs R&D and testing to advance aeronautical and space technologies. NASA's 2022 budget was estimated to exceed $13 billion.

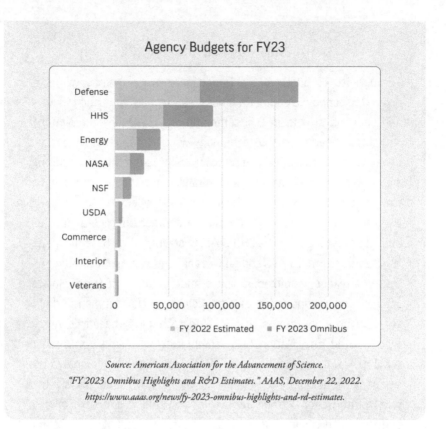

Agency Budgets for FY23

Source: American Association for the Advancement of Science.
"FY 2023 Omnibus Highlights and R&D Estimates." AAAS, December 22, 2022.
https://www.aaas.org/news/fy-2023-omnibus-highlights-and-rd-estimates.

It is important to note that investments in a particular area—the environment is a good example—are usually not confined to a single agency. Therefore, when developing a strategy for engaging with the Federal government, it is important to think broadly about all agencies engaged in work that may be of interest or a match for possible partnerships. Among others, solar technologies frequently have applications across multiple agency missions, such as defense, energy, and agriculture.

In addition to understanding more about primary departments and agencies within the Federal government (and their histories, as I noted in chapter 2), I recommend keeping abreast of what are considered to be priority technology areas. Especially as we look to the future, the Federal government is keen to ensure that the nation invests in areas that help the nation to create or maintain a leadership position.

TECHNOLOGIES OF THE FUTURE

MICROELECTRONICS

Today, microelectronics are ubiquitous across all domains, from defense and national security to education, healthcare, communications, transportation, and the economy. Digital technologies played an enormous role during the pandemic, allowing dramatic shifts to working from home, online commerce, and vaccine production and delivery.

NANOTECHNOLOGY

These technologies focus on the development of new materials at the atomic scale, offering the potential for breakthroughs across many fields, from medicine to computing to energy to environmental protection. As discussed previously, nanotechnology applications today already are pervasive in everyday products and nano-based products are poised to have an even more dramatic impact on our economy, national security, and society.

BIOTECHNOLOGY

Biotechnology utilizes cellular and biomolecular processes to create breakthrough products and technologies with a wide array of applications, such as treating illness, improving the environment, growing the food supply, promoting energy efficiency, and improving manufacturing processes.

QUANTUM COMPUTING

Combining the power of mathematics, computer science, engineering, and physical sciences has the power to change the game in medicine, encryption, chemistry, development of new materials, and more. Quantum computers have the capacity to solve much more complex problems in a fraction of the time required by standard computing.

ARTIFICIAL INTELLIGENCE

AI focuses on the simulation of human intelligence processes by machines, especially computer systems. Applications of AI include expert systems, natural language processing, speech recognition, and machine vision.

Generative AI, as shown by Chat GPT, and similar programs, offers tremendous potential for human-like interactions, offering the potential for enormous gains in productivity, and a complete transformation of the way Americans work, as well as potentially offering a dark side. Many leading innovators have actually called for government study and potential regulation of this powerful technology.

HOW FUNDING LEVELS FOR R&D ARE DETERMINED

When I was in government, one of the roles of my agency, the Office of Technology Policy, was helping to formulate and advocate for government programs of interest to the President and the Administration. An explicit part of President Clinton's technology strategy involved expanding the Advanced Technology Program, a program that had gotten its start, along with a relatively small amount of funding, during the George H. W. Bush Administration. The attitude on Capitol Hill was quite contrary to Clinton's vision, so our agency leadership, my colleagues and I spent months talking with individual Members of Congress encouraging them to do two things: first, not eliminate the program, and second, invest in its growth. Our first big accomplishment was that the program was not eliminated; our second was that it grew for some time thereafter.

Another key role of the agency was to promote competitiveness, so we spent a great deal of time listening to the views of industry representatives and trying to reflect those views in the policy-making process. My colleagues and I traveled all over the country to large cities and small towns meeting with companies, nonprofit organizations, universities, and elected representatives and then turning their input into recommendations for Presidential and agency plans as well as an annual budget. The tech community brought forward a wealth of ideas, everything from streamlining red tape to suggesting specific types of R&D investments that would be beneficial to pro-innovation tax and regulatory policies. That is just one of several ways that private sector and government partners contribute to the policy-making process for Federal departments and agencies.

For those seeking to work with government, it is important to have a basic understanding of the funding process and to take into consideration that the agencies with which you may want to work may have boundaries on being able to enter into new relationships during certain periods. Of course, you can meet with agency representatives anyway and plan for joint initiatives to come.

The process of funding Federal R&D is somewhat complicated. First, the President creates and announces his (or someday her) budget priorities. This will be very detailed, offering line-by-line proposals for all spending, including R&D. This proposed budget then moves to Congress.

Members of Congress review and consider the President's budget. They may or may not agree with ideas the President has proposed. When a proposed budget runs into resistance, there may be quite a lot of back-and-forth between the executive and legislative branches. Eventually, legislators pass two kinds of bills relevant to agency funding (though not for every agency every year).

1. An *authorization* bill, which tells agencies what they can do, and in many cases, what they cannot do. There is no money

associated with an authorization bill other than a theoretical, or proposed, amount.

2. An *appropriations* bill, which provides funds for the authorized agency. As you might imagine, appropriations bills frequently, but not always, follow authorization bills.

In theory, all of this effort occurs at an orderly pace, and in ample time before the beginning of each fiscal year (which begins in October). In reality, there is much delay, putting aside the regular order of things, and racing to the finish line most years. And since 1997, Congress has never passed more than a third of its annual appropriations bills on time.[26] In those cases, Congress typically passes a Continuing Resolution to keep the government operating at the prior year's funding levels; occasionally, the Federal government shuts down due to the lack of agreement on the Federal budget or parts of it. This poses a challenge for all of government, but in the science and technology arena, these delays can be devastating. Under a Continuing Resolution, agencies may not usually start new projects. That situation can put a lot of good ideas on hold and greatly frustrate those seeking to work with the government.

At a minimum, there is great value in looking at a couple of years' worth of authorization bills to see what direction a potential partner agency has gone in (or at least where Congress has directed the agency to go) and then assess whether that direction will potentially help or hinder your attempts to partner with that agency. It is also vitally important to review the relevant appropriations bills to see exactly how much money has been and is being directed to which programs and projects. Appropriations bills may also include directive language, which can give important clues as to what expectations are going to be.

---

26   Drew DeSilver, "Congress Has Long Struggled to Pass Spending Bills on Time," Pew Research Center, July 28, 2020. https://www.pewresearch.org/fact-tank/2018/01/16/Congress-has-long-struggled-to-pass-spending-bills-on-time/.

Every executive branch department or agency is governed by the authorization and appropriations bills negotiated and passed by Congress. These bills are easy to find on the internet, as most Congressional committees and executive agencies maintain useful websites where this information is accessible. It is a good idea to follow the progression of how the bills in which you are interested develop from initial proposals by the President through to the final act passed by Congress and signed by the President. This will give you a good idea of the nuanced discussions that frequently take place and how the process moves forward.

One interesting practice to note is that when an authorization or appropriations bill gains momentum in either the House or Senate, it becomes a vehicle to carry forward other authorization and appropriations bills that previously were stand-alone bills. In the appropriations arena, Congress frequently passes an Omnibus bill when it cannot or does not pass the appropriations bills individually. A recent interesting example that combines authorizing language previously contained in numerous other bills is Creating Helpful Incentives to Promote Semiconductors Act (CHIPS), which became law in August 2022. The bill appropriates over $50 billion for semiconductor R&D and facilities expansion, manufacturing incentives to encourage United States–based semiconductor development and production. The authorization components of the act focus on catalyzing regional and economic development by creating regional innovation and technology hubs, supporting and catalyzing clean energy, increasing opportunities in STEM for underserved groups and communities, and expanding the geographic and institutional diversity of research institutions (including initiatives to support Historically Black Colleges and Universities and other Minority Serving Institutions) and other underrepresented groups.

I recognize that every innovator's situation is unique, but let me share with you some additional ways of thinking about your work that can help determine the fit between your projects and the Federal government. Keep in mind that most agencies care about gleaning helpful data and moving technologies forward. In many cases, agencies will seed numerous ideas and then downselect as some projects prove more successful.

Based on my experience, there are at least five common types of projects sought by various departments and agencies.

TYPE 1: HIGH RISK AND HIGH REWARD

A few agencies, including the Defense Advanced Research Projects Agency (DARPA), the Advanced Research Project Agency-Energy (ARPA-E), and the new ARPA-H (for health), seek breakout ideas that can be game changers if successful. They expect to make some investments that won't succeed, because the projects they support are usually very high risk, both technically and financially.

If you want to apply successfully to these programs, you will need to demonstrate that you have a concrete plan for solving a thorny technical challenge, and if successful, the impact of your work will be very high; this requires a showing of the quantitative risks and potential benefits. You also need to demonstrate that you have a complete grasp of the existing state of technology in your field, state clearly the problem your technology seeks to solve and the potential economic and social impacts of that solution, and show you can think creatively about potential obstacles and their resolution.

TYPE 2: BASIC OR FUNDAMENTAL RESEARCH

Basic or fundamental research is typically considered to be driven by an individual scientist's curiosity or interest in a particular scientific question. The goal is to gain more complete knowledge of a subject without regard to specific applications and with no specific products in mind. Exploratory work of this sort could be done by an academic research team, a national lab, a small or large business, or a consortium.

If you are at an early stage of work—doing a laboratory-based experiment, not yet having won a funding award, and where you have not yet formulated specific questions but seek to increase the nation's knowledge base on a particular subject—your research will fit into this category. There are numerous options for advancing your work, so it is important to look below the surface of program information to determine whether you know the program manager by reputation or understand other projects that the program has supported. Are the program managers people you would enjoy working with or would like to learn from? If the answer is yes, this category is likely for you.

TYPE 3: APPLIED RESEARCH PROGRAMS

In contrast to basic or fundamental research, applied research seeks a solution to a specific problem. Applied research projects are, by definition, closer to market, and some Federal programs specifically target only applied research. Programs run by the Department of Energy's Office of Energy Efficiency and Renewable Energy are specifically geared toward closer-to-market opportunities that can contribute to transitioning the nation's energy supply to net-zero emissions solutions by no later than 2050. When you take the applied research route, understand that government's goal is to solve defined problems in a defined amount of time—often within a time horizon of three to five years. Projects supported will still be sufficiently risky and uncertain such that many organizations cannot invest on their own. It is very important for you as an applicant to be able to quantify the potential benefits of your projects and demonstrate significant impact if your projects are successful.

Many agencies publish strategic plans for their applied R&D agenda, and frequently, agencies will publish specific performance targets.

For example, when it comes to national defense, many of the defense agencies publish *long-term* strategic plans, such as those illustrating the capabilities a warfighter will need to maintain an edge on the battlefield. One example of a long-term development project is the Army's Next-Generation Combat Vehicle, which is expected to operate in both manned and robotic modes and be coordinated with very sophisticated drones and communications equipment. For the Next Generation Combat Vehicle Program, the Army released very specific performance targets for a wide variety of technologies and components anticipated to be incorporated into the vehicle. Needless to say, there is yet a lot of work to be done to make good on the goals for this vehicle. Several teams competed for this major project—which requires both technology development and demonstration—and a six-member consortium, led by Lockheed Martin, won the contract.

TYPE 4: SMALL BUSINESS

Programs such as the Small Business Innovation Research Program (SBIR) are specifically designed to help businesses with fewer than five hundred employees. Eleven Federal R&D funding agencies run

SBIR programs, typically with one or two annual competitions. SBIR funding operates in two phases. Phase I awards are usually under $300,000, but Phase II awards can be much larger. Some agencies, such as the NSF, also offer a supplemental award, available without additional competition.

SBIR has historically been a vitally important resource for qualifying companies. TechVision21 has helped numerous companies with SBIR projects. We worked with a small company that developed an innovative material that could absorb (and remove) oil, pesticides, and other toxins from water. The CEO of this company built a strong, long-term relationship with a single program manager at NSF who had a great interest in the technology. Although working only with the SBIR, this company eventually received several million dollars from NSF, funds they needed to bring their technology to fruition. Today, the company is marketing and selling this product for sampling and capturing environmental contaminants and has expanded into personal care products with applications in foundations and powders, degreasing, extracting essential oils, and dry shampoo, to name a few.

It is relatively easy to qualify for SBIR; the program supports a wide variety of businesses, including women-owned and disadvantaged businesses, and a wide variety of types of research projects. It is highly competitive, so like other programs, it is important to have an innovative project concept and to prepare a quality application. In addition, SBIR can be a stepping stone to other Federal funding opportunities, including direct sales to Federal agencies for those that are ready, subcontracting opportunities, or other avenues for growth and success. Finally, winning an SBIR award can function as a "Good Housekeeping Seal of Approval," providing credibility that then helps a company capitalize on other opportunities.

TYPE 5: DEMONSTRATION

To commercialize innovative technologies, it is often necessary to demonstrate their effectiveness in a real-world environment. There is a huge difference between demonstrating success in the laboratory and actually commercializing a product. Depending on the technology (or system), this can be a costly endeavor.

One historic weakness in the American Federal funding system for innovation has been a failure to invest sufficient amounts in demonstrating and deploying promising technologies. The United States long pursued an approach of investing only in the front-end R&D and then assuming the industry would step in and do the rest. That worked well in the postwar decades when American industry was unmatched in its capabilities. Today, now that the playing field is becoming more even, one major national goal should be to increase the likelihood that technologies developed in America are deployed here, particularly when funded—in whole or in part—by our taxpayers.

In my experience, even entrepreneurs working within large and wealthy companies cannot always obtain the corporate support and financing necessary to demonstrate a new technology or system. Demonstration costs can easily exceed several million, or even tens of millions of dollars. However, opportunities to seek funding for demonstration have increased dramatically, particularly at the Departments of Defense and Energy. This should be a welcome change for American innovators. If you have projects that require demonstration before commercialization, you should investigate where opportunities for demonstration support exist. To do this, you will need to first identify programs that support your field of technology and then research within those programs to learn whether they support demonstration projects, and if so, of what type and scale.

STAY ROOTED IN THE PRESENT

When considering partnering with government, some innovators focus on which political party is in power. They think to themselves, "I'll wait until *my* party gets into office and then make my approach." But this attitude does not take into account the fact that most government programs have multiyear plans spanning multiple Administrations. It is true that priorities and policies do change from one Administration to the next, but there is a large cadre of great people working in senior career positions who remain a constant driving force within an agency, no matter which party or people are in power. If you think you have a technology with economic or societal impact, my advice is to forge ahead. When the political situation changes—and it always does eventually—those changes may make opportunities for you more or less desirable. There is no time like the present.

That said, many current laws, regulations, and financing mechanisms are geared toward or reflective of incumbents. This means that when you want to introduce a new technology, you need to ensure the vision you paint matches current needs in government. It is important to be able to envision what the future might bring—after all, an innovator's job is to stay ahead of the game—but even projections and speculations must be rooted in the present. There is no substitute for observing trends, taking note of which projects receive funding, talking with current program managers, being responsive to current needs, being aware of the competitive environment, and including current data to support your claims.

Working with the Federal government requires patience and persistence and both short- and long-term strategies for building strong relationships.

Nearly every successful government-assisted project I have seen included an offering that was either brand new to the world or a groundbreaking idea that promised dramatic change. Once you have determined that you have a revolutionary concept that is suited to meeting a national need and that partnering with the Federal government is right for your organization, you will begin matching the specific projects you are engaged in, or have planned, with specific government agencies, programs, and funding opportunities. As you will see in the coming chapters, working with the Federal government requires patience and persistence and both short- and long-term strategies for building strong relationships. This work is a lot of fun, as you are likely to discover connections and opportunities that were not obvious to you at the start of your efforts.

I shared a story in chapter 1, one detail of which brushes against the grain of what I would normally advise. That was the story of a small company with a CEO who developed a close and productive friendship with an NSF program officer. I recognize that it is an amazing circumstance to have a good friend who wants to support your work and does just that. But that is an exceptional situation. Given that none of us can count on developing a close friendship with a government employee who is deeply interested in funding our work, I maintain my advice that the best thing you can do is widen your net to include multiple agencies and professional contacts who might be in position to support your goals.

TAKE ACTION

Take a moment to note your responses to the following prompts:

1. Ask yourself this basic question: In what ways is my project unique/state of the art/innovative?

2. Identify ways that your project clearly intersects with the overall government aim of serving the nation.

3. Consider which agency and program missions your project best supports. In most cases, this will result in identifying multiple agencies and programs as possible targets for interaction.

4. Identify which type of project your innovation fits into. Are you a "high-risk and high-reward," "early-stage research," "small business," "applied research," or "demonstration" fit? Something else?

If we knew what we were doing, it would not be called research, would it?

—ALBERT EINSTEIN

CHAPTER 4

STRATEGIZE YOUR FIRST MOVE

IN PRIOR CHAPTERS, we reviewed the importance of taking a thoughtful approach to determine whether your organization and its projects might be a fit for Federal government departments and agencies and the value of bringing knowledge of historical and functional aspects of government processes to bear as you explore partnership opportunities. Here, we take the next logical step that precedes contacting any of the specific agencies or programs in which you are interested. This involves identifying the list of potential government agency partners your project is most likely to fit and then researching those agencies' missions, histories, priorities, goals, and funding profiles, and the current status of your business or project.

In other words, strategizing your first move toward making more formal contact with one or more government programs requires yet another layer of assessment.

DO YOUR HOMEWORK: GET YOUR DUCKS IN A ROW

If you want to make an accurate assessment about which agencies or programs are likely a good match for your project, consider generating thorough answers to the following questions:

EXACTLY WHAT IS THE NEW CAPABILITY, PRODUCT, OR SERVICE THAT YOU ARE OFFERING?

Consider that you likely will need to educate your audience about the value of your project, product, or service. In my experience, most entrepreneurs know where they are copying what others have done or are doing and where, or in what ways, what they offer is unique. If you are unsure about the status of similar projects in your field, one resource to which you can turn is the Patent and Trademark Office. Carefully researching and identifying other organizations claiming to work in your same space will allow you to address them all one by one. You should know where your product or service stands in relation to its closest competitors, if any. The best case for your uniqueness will always include an accurate assessment of where and how what you are developing intersects with and offers some superior performance characteristics to what others are doing.

It is important to make your honest best case for the novelty and potential impact of your offering. Consider what a government department or agency would need to know about the status of innovations within your field and the way your work produces something new or otherwise transforms known approaches to a problem.

Not only do you need to clarify the precise way in which what you have to offer is new, but you need to be able to describe it so that it makes sense to both highly skilled technical audiences and intelligent nonspecialists. Use a brief set of slides with a simple message and/or a brief paper. You need to be able to make all key points in a few paragraphs or, at most, a few pages, and write in plain English, avoiding unnecessary jargon.

WHAT ARE THE STEPS INVOLVED IN CREATING THAT CAPABILITY, PRODUCT, OR SERVICE?

As we reviewed earlier, certain categories of funding opportunity are geared toward specific points in the R&D and commercialization cycle. Determine which of your projects are in which stages of development. Frequently, companies pursue multiple categories of funding opportunity at once, given the status of different projects—a commercial product ready to sell; a near-commercial project for which they are engaged in demonstration, testing, and evaluation; and projects in earlier stages of research and development may all intersect with government's goals and needs, though they would compete in different categories of funding opportunity or pursue different kinds of government partnerships.

When it comes to communicating about the steps in your process, focus your story on the way what you do improves upon the state of the art. A handful of well-edited sentences and relevant corresponding data should suffice. My advice is that you think of this less as a review of everything you've done in the past decade or every single detail of your process and more as an opportunity to shape the narrative around what you have to offer. The way you choose to present your work is yet another opportunity to educate and persuade. To the extent that you are able, try to anticipate audience biases or presumptions about prior work and the current status of your field. The more you know about government agency and government policy history, the more likely you are to be able to respond wisely to assumptions that may no longer be grounded in current data.

HOW MUCH TIME WILL IT TAKE YOU TO ADVANCE FROM YOUR CURRENT POINT IN DEVELOPMENT TO ACTUALLY HAVING A PRODUCT, SERVICE, OR OTHER CAPABILITY THAT YOU CAN DELIVER TO THE MARKET?

It is one thing to estimate the overall timeline to completion and another to account accurately for all the milestones that need to be reached and tasks that need to be accomplished to get to the point where you can commercialize your product.

Be honest in your estimates and heed the maxim: everything takes longer than you expect. Many companies are eager to get their products to market and overlook opportunities for government partnership on the assumption that doing so will only slow a process from which they hope to profit. This is not always the case. There are times when government partnership has accelerated the development of a specific technology, and there are times when companies would not be able to move projects forward in the absence of government support. Electric vehicles are a good example of this; government support was instrumental in moving EVs from novelties to the market. But even though mass-market EVs have been available for over a decade, they still represented under 5 percent of new car sales in the United States, as of September 2022.[27]

Be honest in your estimates and heed the maxim: everything takes longer than you expect.

---

27 Sebastian Blanco, "Electric Cars' Turning Point May Be Happening as U.S. Sales Numbers Start Climb," Car and Driver, September 7, 2022. https://www.caranddriver.com/news/a39998609/electric-car-sales-usa/.

WHAT RESOURCES DO YOU HAVE AT HAND, AND WHICH DO YOU STILL NEED TO ACQUIRE?

Consider this less a question about money and more a question of access and expertise. Bringing a new product to market does take money (sometimes a lot of it), but it also takes the right kind of skills and talent, equipment, and facilities. What kinds of entities are you competing with in this space, and what resources do they have? Is the team that brought you to this point the right team to carry you forward through the next phase? Do you have a team of technologists working in the laboratory who *also* have experience taking firms to the next level? What kind of workers will you need to manufacture your product? Other considerations include knowing the kind of facility and equipment you will need and the specific partners you may need—which could include a Federal laboratory or other agency that supports user facilities.

For example, at one end of the resource spectrum, there is software production, which in most cases is relatively cheap. You need people, computers, storage, and similar resources. At the other end of the spectrum, there could be a production process requiring a factory floor full of manufacturing equipment; depending on the equipment and the industry, some of that equipment can sell for multimillions of dollars. If you are a small company and you have reached your current position by being creative, having good ideas, and not spending much along the way until you arrive at the production hurdle, keep in mind that the answer to clearing this hurdle may not always involve building your own factory. Instead, you might need to rent space or utilize another company's resources by partnering with an existing factory setup that matches your need.

Part of making these decisions about production has to do with the price tag and the complexity of the equipment needed. The semiconductor industry has extremely high production costs—with individual pieces of manufacturing equipment selling for tens to hundreds of millions of dollars—and semiconductor fabs must be kept current and ready to create the most sophisticated product. For some, that may be best accomplished by using existing facilities rather than building them oneself.

Another potential consideration is the competitive landscape. Sometimes, the right partner in terms of its facility and equipment offerings may also be a direct or quasi-competitor. It may be that the need to keep your business and production processes top secret limits your flexibility when it comes to engaging with available resources in your field.

What I have noticed from my years of experience working for and with government funding agencies in technology is that there are people who begin with a well-planned-out strategy and understand where they are today, where they need to go, and what steps they need to take, including investments in equipment and partnerships. But others, including many small companies contending with their first innovation and where the founder or founders have not had prior experience with all parts of the process, can find themselves with a great idea, a prototype of their project, and no business plan whatsoever.

The recommendation I would make here is one a reader is likely to expect. Putting together a plan early in your own R&D process is best. I have seen some of the smartest people in the world come up with new and fantastic ideas. But what they know best and most is *not* how to go about the business of making the product they envision. There is another set of skills required for making the transition to market. If you do not have experience (or do not have staff with experience) taking a product from its inception in the lab all the way to the marketplace, do bring on someone who has quantifiable expertise successfully guiding that process to completion. Eventually, you will need to acquire management and financial expertise, marketing capacity, a high-quality and easily understandable website, and someone on board who knows how to sell and close deals.

Once you have identified your plan for taking your company from idea to market, you are ready for the next step.

READ THE TEA LEAVES

It is important to look at as many relevant Federal government sources of information as possible. Most agencies' websites reveal detailed information about their budgets, Congressional testimonies, programs, funded projects, and even future plans. In addition, these same agencies often publish technology road maps and hold program reviews and conferences to gather and share information. Before you contact any program or agency representatives, you want to have a good sense of where the agency fits into the overall portfolio of Federal agencies, what the agency has accomplished to this point, what it is doing now, and given its current status, where it is headed in the coming years.

Be sure to review the following:

AGENCY MISSION AND HISTORY

An agency with a hundred years of history will be different from a newly created agency in many ways. With the former, programs are well established, application requirements have been stable for many years and are usually easily accessible, and prior winners are known.

FUNDING HISTORY

How much funding has Congress given the agency or program? What funds were appropriated over the last few years, and are there any signs of where agency funding is headed? For example, if last year the program received $100 million and they are proposed for $150 million in the coming year, it's fair to assume that they are in a growth cycle. If they got $100 million last year and they are going back to $50 million in the next, that organization may be undergoing reorganization or even going out of business.

FUNDING OUTLOOK

What program materials can you locate that indicate the agency's priorities going forward? Has it published any program plans or reports? Has the head of the agency given a recent speech in which priorities are actually named? Have agency officials recently testified, been called to answer questions before Congress, or submitted any written materials to Congress indicating the organization's agenda? Learn as best you can where the agency intends to place its focus going forward, at least in the coming twelve to twenty-four months. Putting together insights drawn from stakeholder meetings to program plans, Congressional testimony, and budget documents will give you the fullest picture of an organization's outlook.

HISTORY OF PROJECTS FUNDED

As you look through the list of past and currently funded projects, consider what threads them together. Do they speak to a particular set of needs? Do they approach those needs in a particular way? Are they focused on modernizing existing technologies or in pursuit of alternative technologies? If there are projects that stand out, in what way(s) are they both similar to and different from the rest? Being able to categorize these projects can give you a strong sense of how your own work fits, or *could* fit, into agency funding patterns.

NETWORK CONNECTIONS

If it is obvious which people are in charge of a program or agency, consider whether you might have some connection to one or more of them. These may be people to whom you reach out initially, whether by talking with them at a conference or contacting them directly to discuss the compatibility of your work with the program's aims. Work on identifying the people it makes the most sense to talk to, along with the people you would *like* to talk to. Is there a person at the agency who works in your same area? Put that person on a list of people to contact.

> *In my experience, the most common self-imposed limitation an organization can choose in this phase of preparation is to look only at the departments and agencies that are the most obvious matches for their work.*

In my experience, the most common self-imposed limitation an organization can choose in this phase of preparation is to look only at the departments and agencies that are the most obvious matches for its work. It is good to identify the obvious matches, so long as your research does not stop after having done that. If you have a product that is going to reduce the cost of electricity, it makes sense to consider taking that product to the Department of Energy. But it would be a shame to stop there and not look into opportunities for partnership with the Departments of Defense, Agriculture, and Commerce.

Thinking beyond very high-level correlations can open up additional opportunities, *including ones that may even turn out to be a "best fit" for your work.* Earlier, I noted that because of its wide-ranging set of missions, it makes sense to check Department of Defense websites for possible connections on almost any project. Here, I will add that this same advice applies to the Department of Agriculture.

POLICY ENVIRONMENT

It can help to think through the overall policy environment starting with the highest levels of government. An incoming President looks at what the government is doing already, which is carrying out the agenda of the prior Administration and abiding by its funding levels. A new Administration might choose to build on existing capacity, or it may choose to move in a different direction. In either case, any new Administration takes over the budgetary process of the prior one midcycle, so any changes decided upon by a new Administration will not take effect until the following fiscal year—at the earliest. There is more detail to the policy-development and appropriations process, but I mention it here to suggest that it is worth considering the rhythm of government policy-development cycles as you are putting together your strategy.

All that said, I offer a general word of caution about public information: What an agency advertises to the public should not be assumed to be the full scope of its work. Often website-level information tells a small part of a much larger or deeper story, and it may do so in a way that speaks so broadly that determining the nitty-gritty of compatibility is not visible there. You may have to dig a bit deeper for more specific information. Of course, once you get to the point of meeting and talking with agency representatives, you will be able to glean from those conversations more detail about fit.

Even with a lot of preliminary strategizing and preparation, it can sometimes turn out that there is a mismatch. However, by the time you determine that this is the case, you will likely be in conversation with someone from the agency who, given their insider knowledge, can refer you to another person or program with whom you can continue the dialogue. One contact leads to another, one bit of research to another. The good reasons for engaging in preliminary research and strategizing are to get a much stronger sense of the multiple ways you might move your project forward and to be able to engage in conversation with program leadership in ways that indicate you have done your homework.

ENGAGE IN MULTIPLE WAYS

Once you have determined that you may have a fit somewhere, look for opportunities to pursue contacts there. Perhaps an agency with which you hope to partner is hosting a conference, or perhaps several of its representatives are giving presentations at another conference or meeting. Attend those gatherings and look for opportunities to chat with those people. Meeting people at events and striking up a casual conversation about agency priorities and how your work speaks to one or more of those priorities is a great way to engage before trying to schedule a more formal meeting. Of course, scheduling that formal meeting is a necessary part of your plan, but it need not be the very first thing you do. Likewise, you may be able to arrange a meeting alongside the event or conference rather than taking a chance on striking up a conversation in a crowd.

I suggest that even when making an initial, more informal approach, you prepare some materials that can be left behind. That may be a handful of slides, a brief white paper, or some other succinct project description. The key to these materials is to keep them brief but highly informative. As we discussed earlier, your materials should clearly explain what you have to offer that has not been seen before. The worst thing that could happen is that program managers leave an interaction with you thinking that your project is just like that of the last hundred people with whom they have talked.

If your plan and timeline do not allow you to pursue an initial, more informal conversation, consider sending those brief, focused materials directly to an agency representative. In so doing, you can initiate contact, get an early read on the representative's interest, and propose setting up a meeting for a longer, more formal presentation of your work.

Essentially, thorough preparation allows you to be the best salesperson for your project. Know where your work fits within the field, create a one-to-two-page document outlining just that, and be ready to explain and defend why your approach is superior to what others are doing. A clear message along with some humility—both about what you are working on and about what you know about the program you are seeking to partner with—can take you a long way toward establishing a productive relationship with a government agency.

TAKE ACTION

1. Ask yourself: What resources does my business need to move forward from its current position in the R&D cycle?

2. Review each agency's mission and history, the policy and program environment, and the funding history and outlook. To optimize results, look across multiple agencies where your projects might be a good fit.

3. Identify potential contacts and connections at agencies, given those contacts' project portfolios and stated interests.

4. Consider strategies for meeting and establishing connections with the people you identify.

If you can't explain something simply, you don't understand it well. Most of the fundamental ideas of science are essentially simple, and may, as a rule, be expressed in a language comprehensible to everyone. Everything should be as simple as it can be, yet no simpler.

—ALBERT EINSTEIN

CHAPTER 5

CRAFT AND DELIVER YOUR MESSAGE

AMONG MY RESPONSIBILITIES during my eight years working at the Commerce Department was advising on creating programs and specific funding processes to determine and apply rules by which entities could qualify to compete. That process, I will note, is less flexible overall than the policy-making process. There were some programs that made room for revolutionary new products and processes and encouraged applicants to take risks; other programs were geared toward meeting a clearly defined set of specific goals and required applicants to demonstrate their capacity to knock those goals out of the park in order to receive funding.

While working at the Department of Commerce, I had the opportunity to partner with representatives of the Departments of Energy, Education, Labor, Agriculture, Defense, and NSF on a regular basis. Knowing the lay of the land within and across the missions of these agencies allowed me to help those curious about partnering with government to determine where they might be a good fit.

At the Department of Commerce, we had a stream of people come through every day to present their work—many of them genuinely unsure where their projects belonged. For some, this was because their projects sat at the intersection of different departments and sets of goals. For others, this was because they had yet to think carefully or had limiting preconceptions about fit. For all, it was my pleasure to hear about their work and help them make contacts across agencies that would get them one step closer to establishing a flourishing partnership. In some cases, it was also my goal to encourage them to take a more holistic approach, to broaden their vision of how the technological innovation they brought to the table could contribute to improving the national economy and national security and meeting other national goals.

I must have listened to hundreds of presentations from organizations seeking Federal funding and partnerships for a wide variety of technologies. I learned a lot from these presentations and had the opportunity to support some of the technologies on which I was briefed. One of the things I valued most during my time in government was having people come in to visit who were well organized and brought with them concise materials and a clear message. I have carried those principles forward as I now help innovators prepare to brief and work with government.

We will focus in this chapter primarily on the presentation of your projects during arranged meetings with government agencies and programs. But before we do, let me mention again the great value in also pursuing more informal opportunities for connecting with government officials, including joining relevant professional or industry organizations or attending conferences, think tank meetings, and other presentations by guest speakers in your areas of interest.

You should take advantage of all opportunities you can identify for informal engagement.

The overarching goal is to establish yourself in a community of people who share similar or relevant interests. Doing so creates more informal opportunities for conversation and learning—even opportunities to brief people at a given agency or organization with whom you may not meet when it comes time to give a more formal presentation of your work. In a later chapter, I will address more directly the importance of briefing people at all levels of an organization and across organizations and levels of government. Here, as there, my point is to take appropriate advantage of opportunities to share what you are working on in ways that establish your keen understanding of the areas in which your work is focused.

To that end, formally scheduled meetings are invaluable and may be a one-time opportunity to explain your organization's mission, goals, specific technical projects, and the outcomes you seek from working with government.

ARRANGE PEER-TO-PEER MEETINGS

I recommend that you choose which people within your organization will attend a meeting based on the person or people with whom you are set to meet at a particular agency or program. The goal is to bring to the meeting the position counterparts or "peers" of the people you will be interacting with. In other words, if your organization has set up a meeting with a very senior Senate-confirmed appointee and that person's staff, you would want to ensure that someone with a very senior position in your company or organization attends the conversation.

Similarly, if you are meeting with an agency or program representative who is not a senior-level person, consider not inviting the most senior-level person in your organization to attend. Likewise, if you have someone who is particularly good at answering highly technical questions but can come across as awkward to nontechnical audiences, consider having that person sit quietly through the overall presentation and only speak to the most technical aspects of the project if and when anyone asks about them. Alternatively, that person may be best utilized when it comes to answering questions in a follow-up correspondence or other written document.

In an earlier chapter, we addressed the importance of making contact with individual agency representatives whose interests appear well-aligned with those of your business or project. Here, my additional recommendation is to take note of those people's seniority levels and do your best to match that with the appropriate level of seniority within your own organization. If you have already been attending conferences and meeting relevant contacts along the way, you will know that this process of establishing peer and personal connections often occurs quite naturally and easily in that environment. Maybe you approached someone because you were intrigued by the question they posed during a Q&A, or maybe you appreciated and followed up on some element of a speech or presentation they gave. A meeting could end up being proposed in those interactions—"May I come in and talk with you? I think you'd be interested in hearing what we're working on."—and make easier work of determining who else might need to be involved.

CUSTOMIZE YOUR PRESENTATION

One of the most important things you can do to prepare for talking with agency representatives is to consider and customize your presentation for your audience *each time you schedule a meeting with particular individuals or groups.* One way of doing this is to consider:

- The audience's technical expertise.
    - Will you be speaking with technical experts?
    - A nontechnical audience?
    - A group of people that includes both?
- The stated interests of the individuals with whom you will be conversing, if this can be discerned.
    - What is their background and training?
    - What program(s) have they run in the past?
    - What are their passion projects?

A presentation tailored to address the interests and commitments of the program and of the people in the room with you is one way of demonstrating that you have done the preliminary work of assessing the compatibility of your project with their goals.

In my experience, the most challenging aspect of preparing and presenting one's projects to government audiences is bridging the gap between technical and nontechnical expertise, particularly when technical experts are invited to speak with nontechnical but expert audiences.

To be clear, in all cases, it is imperative to offer enough detail and explanation to be convincing about the current and future directions of your work, what you are trying to accomplish, and the kinds of contacts that you are hoping to make. But an appropriate presentation includes only as much detail and jargon as is necessary—and no more. In many cases, more comprehensive and more technical information can be held back from an initial presentation and put to good use when responding to specific questions or requests for additional detail.

As you might imagine, there will be situations in which an agency representative will have a highly technical question that requires a highly technical answer. The key to a good meeting—and to a good relationship overall—is to be able to move elegantly between less technical and highly technical commentary, as the subject matter and the conversation demand.

If an agency representative asks a business question, one intended to determine market credibility or your organizational strategy, be sure to answer precisely that question and not take the opportunity to forge ahead with a presentation or to fit in a bit of detail about one of the more technical aspects of the project. Business questions—as much as technical questions—help agency representatives assess your overall competence and readiness and can help them answer the question, Is this someone with whom we would like to work?

FOCUS ON THE ESSENTIALS

Anyone proposing their project to the government should be prepared to highlight clear answers to the following two questions:

- Who are we?
- Why are we here?

as well as seek answers to the following two questions:

- What kind of partnership are you seeking?
- What is the best way to work with your organization?

I remain surprised by how many presentations I have heard that they did not do a successful job of addressing these essential criteria for introducing yourself to a government agency. Instead, many begin with the assumption that the agency people already know the value of the work that they are doing, instead of understanding that the goal of the meeting is to make a convincing case for the value of what you have to offer. Or alternatively, presenters assume the audience knows nothing and come in armed with fifty slides, which may or not be of interest.

The goal of the meeting is to establish a dialogue, so you do not want to eat up all the time doing all the talking.

It is good to be so prepared for a meeting that you can move in any direction with knowledge and ease. But all that preparation need not be revealed in the conversation itself. Instead of fifty slides, perhaps between ten and fifteen will suit your purpose—even if your project is fairly technical and detailed. The goal of the meeting is to establish a dialogue, so you do not want to eat up all the time doing all the talking. I have noticed that people are so proud of what they are doing that they can talk for a very long time about it. That can be a wonderful thing in certain settings, but when you are presenting your work to the government during a set, and limited, time frame, you need to be very disciplined about your message. Make sure you prioritize and communicate the top handful of messages and leave the audience wanting more. In some cases, detail is a necessary part of imparting understanding to your audience. But as you prepare, you should ask yourself when and whether a particular detail is a necessary component of the story you are telling. Perhaps you are giving too much information where a gloss or even a summary would do.

It is always an option to follow up after the meeting, whether that is by sharing materials you have prepared or gathered ahead of time—a white paper, an additional slide presentation—or by making a plan for near-future contact such as a telephone or video call or even another meeting. Just remember that in following up, the materials you share should further customize your message to known audience interests and, as warranted, invite more conversation.

AIM FOR CONVERSATION

I have seen many large and sophisticated companies bring in a 50-slide presentation and six speakers, then stick to their original meeting plan despite obvious discomfort of the government representatives. I cannot emphasize enough that it is critical to pay attention to your audience's verbal and non-verbal cues. It is not necessary to present every slide your organization has prepared or to give every person attending the meeting a speaking role. Keep in mind that an ideal meeting will proceed much like a good conversation and not like a lecture. Try to establish a personal connection early on. Maybe you went to the same university as your conversation partner or worked at the same company. Maybe you have a network connection in common.

Additionally, you will want to demonstrate that you have done your research by drawing on that research to speak knowledge-ably about the organization or program *at appropriate moments in the interaction.* You may even find a way to use what you know as a means of learning something you do not yet know. For example, if, as part of your presentation or in response to a question from an agency representative, you were to say, "I remember reading in your annual plan that [X] is a priority. Is that still the case? And, if so, could you tell us more about it?" With that question, you can prove that you have done your homework, seek confirmation or correction of what you believe you know already, potentially acquire more detailed or up-to-date information, and, potentially, achieve greater clarity about agency priorities and the actions being taken in response to them.

PREPARE A SET OF BASIC QUESTIONS

One of your aims is to establish rapport, and when you are trying to make conversation, one easy way of engaging the people you are speaking with is to ask them smart questions. It is quite fine to ask questions about what you have already researched prior to the meeting, as long as you do so in a way that shows you have done your homework. As I noted earlier, you are likely to discover what has not been said on a website or in a planned speech. It is OK to ask versions of questions like:

Beyond what you have on the website,

- What are your priorities for the next year to eighteen months?

- Where do you see the agency's work headed?

- What do you think your budget will be in the coming year?

- Do you have specific plans for putting out a funding opportunity?

- We see from your website that you've partnered with this type of project in the past; will you continue in the same direction?

- And, what we've heard from you is very intriguing. What is the best way for us to work with you?

Some agencies may even have an advisory or steering committee with which you can also engage. Always ask rather than assume you know all there is to know.

KEEP TRACK OF TIME AND USE IT WISELY

When you have just thirty or even sixty minutes with a government agency representative, you need to ensure that you accomplish your goals for the meeting on time or even a bit ahead of time. It can often fall to you to direct how time is spent during the meeting. Having a strong sense of what material you need to cover—and by what point in the allotted time frame—can ensure that you succeed in addressing all your primary talking points. That said, some of the best conversations between potential partner organizations and government program representatives occur when the potential partners are prepared for anything that might arise during the meeting. I have seen too many presenters grow flustered by questions they had not prepared to answer or by being asked a question the answer to which might significantly alter their original plan for the presentation.

Visible interest on the part of the government agency or program representatives is a wonderful thing—that includes enthusiastically interrupting a presentation with questions that show they are attentive and engaged and already thinking forward through the material you have presented. If it happens that a program representative asks a deeply engaged question, you and your team need to be able to capitalize on that enthusiasm and address the question immediately—or acknowledge that you are about to address it, even if that means you have to shift your plan and address your points in an order other than the one you had intended.

Let the conversation move forward as naturally as possible, and do not be so bound to your presentation that you end up backtracking or otherwise slowing that momentum. So if the person that you are meeting with wants to take things in a different direction from what you planned, roll with it. I say that because even when you have done some careful research and preparation, it is still possible to be surprised by the direction a program officer takes the information you have to share. Stay engaged with the conversation you are actually having rather than the one you may have previously imagined having, and you may be able to get recommendations that help you move forward with the agency.

On the other hand, if your audience looks uninterested, even bored, or keeps checking watches, phones, or clocks, stay focused on emphasizing your main points, then do your best politely to wrap up the meeting. For example, a presenter might acknowledge, "We know you're a busy person, so let's just focus on a handful of things," and then keep top of mind the primary goal of the meeting: to be clear about why you are there and how you believe your project could address one of the agency's concerns. That meeting might end with someone on your team saying, "Perhaps you can tell us who in your organization you think we should be talking to or working with," and seeing what results. Sometimes, your conversation partners may be proactive about just that, indicating to you early on that you should meet, instead, with one or more of their colleagues.

Usually, if you are the one to ask for the referral, most program representatives will be willing to share the names of people they genuinely believe you should meet. And most of the time, you should not take it personally if your audience looks bored. Sometimes, people are truly very busy or did not know you were on their schedule on a given day because an assistant was the person who placed you there. In awkward events like this, it is important simply to do your best not to offend anyone or further waste time that's not being well spent. You are not obligated, in other words, to stick around to reveal everything you know, nor should you keep trying to convince your audience if there has been a clear sign that the people you are meeting with are not the right match for your work.

Another way of putting this recommendation is as follows:

READ AND RESPOND TO THE ROOM

It is far better to notice that your audience is bored or distracted than to just put your head(s) down and remain unaware, or undaunted by the fact, that things might not be going as well as you had intended. It is better to keep the meeting relatively brief than to continue on at length and at the expense of reminding your audience of your primary messages. I know it can be discouraging to work very hard planning for a meeting and preparing a great presentation only to have a meeting fall flat. If this happens, take a deep breath, try to wrap up by highlighting your best points, and then move on.

Just as you should keep a lagging meeting relatively brief or capitalize on audience engagement, so too should you listen carefully to what your audience reveals in conversation. It could turn out, that the project you thought you were there to pitch seems, after some conversational exchange, not as well matched to the agency or program as another of your projects might be. Or it could turn out that, in real time, you might hit on a way to tweak the project you are there to propose so that it speaks even more directly to an agency or program need that has been uncovered or revealed during your meeting.

It's worth taking your cues from the agency representatives with whom you are there to speak and whose attention you are there, in a sense, to direct. We often do this sort of thing unconsciously, but it is worth consciously noticing details like, Do I have a friendly person here? Is this person indicating to me that they are overwhelmed with meetings or busy with other responsibilities? Is this person displaying interest in our discussion?

It is very important to be sensitive to non-verbal cues; I have seen agency representatives do all kinds of things to indicate disinterest, including looking repeatedly and quite noticeably at the clock, yawning, or flipping ahead in the presentation. Note that skipping ahead could indicate either boredom or interest, so you have to watch carefully. The person could be counting the minutes until the meeting is over or could be so excited that he or she wants to know exactly what's coming later. Even if it seems that a presentation is going well, keep an eye on the clock and don't extend the meeting past the appointed time without a clear invitation to do so.

BE FLEXIBLE

The bottom-line goal, the whole purpose of whatever you say and do, should be to draw out the government representative(s) with whom you are meeting—getting them to clarify their priorities and answer any questions you have, given what you have already learned from your preliminary research. A great hour-long meeting is likely to be one during which the clients spend no more than half an hour explaining who they are and why they are there—what they believe they are accomplishing that is unique and valuable. Then there is half an hour's time that can be spent talking with the agency person about how they are running the program; what's coming up regarding budget requests, funding announcements, and sponsored conferences; and what direction they believe they will be headed in a year's time.

Being flexible applies not only to being open to the many ways the meeting might progress but also to being open to the possibility that there are other projects of yours that may be better fits than you first imagined. So long as you can talk about those other projects with the same preparedness and grace—and not seem to be scrambling to identify anything at all that might apply—you may discover a potential partnership that you had not considered before having this opportunity to talk. Your response to such a discovery might be something like this: "Oh, that's very interesting. We are doing some work in that area …" followed by a brief overview of whatever you are able to report on. Then, "The slides that we prepared for you today don't address this project, but I could send you some follow-up information if you're interested."

EXERCISE PROPER ETIQUETTE

You can meet the goal of positioning yourself as responsive to your audience's interest while also being polite and attentive. That goes for in-person meetings as much as for virtual ones. There is no doubt that virtual meetings make it more difficult to read the room—especially when reviewing a slide presentation—and to assess your audience's body language, but it is just as important to try to adhere to these suggestions on video calls as it is in person. In all cases, consider beginning by thanking your hosts for meeting with you and taking the opportunity to say outright why you believe there is a fit between your project and this particular agency or program. "We wanted to talk with you because your program does [X, Y, or Z thing] that we are interested in, and your background as a solar technology expert aligns very well with the product we've developed and where we think that we're going." Do not forget to offer a clear summary toward the end of the conversation.

In all, and regardless of whether your meeting begins a relationship that eventually yields a partnership, you want the people you have met to remember you favorably. So, no matter what, do make contact afterward to thank your hosts for the opportunity to present your work.

Consider meeting government officials and building relationships to be a journey. You make the best plan that you can, start implementation, and then course correct, as needed. You may end up in an even better position than where you initially imagined.

The final point about meetings is to have appropriate follow-up. Thank-you notes are a must, whether or not you feel that the meeting was a success. Always reach back to say what you learned, and—if things didn't go well or if you arrived at the conclusion that another agency or program might be a better fit, say as much—ask for a recommendation to a person, program, agency, or department that might be a better fit.

In addition, if the agency representative has requested information, send it right away. If no information is requested in the meeting, you could still send a very brief set of information (a couple of slides or a one- or two-page memo). Other follow-up actions will be dictated by the outcome of the meeting and could include the following:

- Meetings with other colleagues or individuals recommended by your meeting host;

- Participating in a conference, workshop, or event that your host agency is joining;

- If the meeting went really well, inviting the host to meet with you at your site, if local, or when traveling in your area. If you have a great show-and-tell opportunity at your site, that's even better; and/or

- If there will be an upcoming funding opportunity, preparing project ideas in advance. For established programs, you may look to prior years' solicitations to give you some idea of the format, information required, and collateral information (such as letters of support).

Regardless of your feelings about a meeting with government representatives, do take the opportunity to debrief after each contact. Together with your team, try to assess matters such as whether and the extent to which you achieved the goal of presenting something truly new and different; what, specifically, can be learned from questions raised or positions expressed in light of your presentation; what amount of time you spent asking relevant questions, listening, and responding to answers; and whether and why moving the conversation in a different direction might have been useful.

The recommendations I have made in this chapter are the product of decades of experience while working as a senior government official responsible for receiving companies soliciting partnerships with government agencies, as well as experience assisting clients of TechVision21. If there is one thing that I can encourage you to keep in mind as you proceed, it's this: a key point of your meeting with government representatives is to convince them of your qualifications to participate in their program or other offering. As you craft your primary messages, think carefully about how you will demonstrate not only the greatness of your idea, process, or product but how it qualifies you as a true government partner.

TAKE ACTION

1. Ask yourself: What evidence do I have to prove the value of my project or product, and on what data and information do I rely to make the case that my innovation has a good chance of working?

2. For each program manager, policymaker, or other government official with whom you meet, identify the top five points that are critical to your overall message.

3. Identify the appropriate people to attend each meeting, taking into consideration their position within your organization and the role they are best suited to play.

4. Manage the meeting, whether it takes place in person or virtually. Watch the time carefully. Make sure to end your presentation within a reasonable time and leave an opportunity for questions and discussion.

To succeed, you need to be resilient. You may have only 30 days to put together a top-notch technical proposal. It's not uncommon for small businesses to spend hundreds of hours and hundreds of thousands of dollars developing proposals ...

—GUIDE TO MARKETING TO DOD[28]

CHAPTER 6

ENGAGE IN THE COMPETITIVE PROCESS

IN THIS CHAPTER, I WANT to focus on one key element of your continued strategy—one that is best pursued in light of these others—and that is engaging in formally organized funding competitions run by the agencies you have identified as having high potential for partnership. Engagement in the competitive process—knowing full well that you are very likely not going to win funding 100 percent of the time—has partly driven my suggestion that you cast a broad enough net when assessing which agencies might be welcoming of your projects. You will want to set priorities among the various programs in which your organization is eligible to compete and think carefully about what resources and time are needed to prepare a competitive proposal and submit it in a timely manner.

From the early point of assembling the right team and drafting sections of the proposal, there are usually several rounds of edits and additions before everyone agrees on the content of the work product

---

28   Guide to Marketing to DOD, Office of Small Business Programs, Department of Defense, accessed May 9, 2023, business.defense.gov/Portals/57/Documents/2017%20Guide%20to%20Marketing%20to%20 DoD.pdf?ver=2017-09-21-121932-073.

and the proposal begins to approach its final formulation. After that, there is the critical need to ensure that the proposal speaks with one voice and provides completely and concisely all the information requested by the agency; this can involve eliminating repetition, adding what's been left out, improving phrasing, and ensuring that the parts form a coherent whole.

At its best, the latter activity typically involves focused work by one or two people with a strong understanding of the organization and its goals, and is capable of integrating everyone's separate contributions into a sense-making, nonrepetitive, no-gap, and convincing whole. That entails, at a minimum, clarifying how the organization is proposing to meet the requirements of the program, what specific results it can anticipate (and why), and both the specific plan (including milestones) for reaching those goals and the specific budget that will facilitate the success of that plan. If you've participated in a proposal-writing effort like the one I describe here, you well know the amount of labor that can—or should—go into producing a competitive document.

TechVision21 has been fortunate to work with many types of companies and other organizations to create funding proposals in response to specific agency calls. For large organizations, it is particularly important to consider people across multiple internal departments—the technical team, R&D, legal, finance, business strategy, and others—to provide content for various parts of the proposal narrative, timeline, and budget. Just pulling together the right team to generate the proposal can be, in itself, a challenge. Then, of course, there is the business of organizing who will contribute to which portions of the proposal, as well as collecting everyone's separate inputs and shaping them into a whole that would be coherent and very clearly responsive to the proposal instructions.

Frequently, companies wait until the time is short to begin working on the tasks I have outlined as contributing to a successful proposal preparation effort. In many cases technical teams are talented, but they had so fully embraced their work that what seemed simple to them—what they believed was clear, at least in theory—required more explanation than they believed was necessary. They often have not yet considered that their reading (and judging) audience would be composed, potentially, of both people who were comfortable using the jargon they used every day as well as people who were not at all familiar with their explanatory shortcuts. And they certainly had not considered that the people who would read their application would be reviewing it among so many others that any company or organization has at being successful relied on its ability to communicate clearly, quickly, and thoroughly. It is true that each application needs to feature what is interesting, unique, or a clear improvement upon other project ideas, but it also needs to be able to back up those claims with targeted explanations, timelines, deliverables, milestones, the right team of researchers, and other indicators that it is on a path to establishing the objectives it has set out to achieve.

TechVision21 has helped many companies with painfully short deadlines. This requires a great deal of extra effort, on the part of both my team and the client. To meet a two or three-week deadline, for example, we have needed to bring in a team of several people to work night and day. After going through this experience, what most people learn is that if you leave all the work to the last minute, you are literally racing the clock to get something turned in, and—if your organization can afford it—you may need to call in a team of professionals to work around the clock to give you your best chance of success.

Especially at the start of your attempts at partnership with the Federal government, *you need to apply to more than one program* in order to succeed within the competitive environment. All that preliminary-seeming effort we discussed in earlier chapters is what will help you make informed decisions about how many and which competitions you will participate in as part of your overall strategy. If, for example, you want to win two competitions within a given cycle or season, a good rule of thumb would be to compete in three or four. The meetings during which you engaged agency or program representatives should help you prioritize your options. Some of the people and organizations you met with likely responded quite positively to your work; others less so. After an honest evaluation, set a plan to focus on the competitions that will maximize your chances of success.

I am going to remake here a point that bears repeating: a fine rule of thumb is to heed the old adage not to put all your eggs in one basket. It is not at all typical for any entrepreneurial effort to sole-source a project through only one government program or agency. The handful of projects that do end up establishing a sole-source relationship are likely to have something widely acknowledged to be unique to a particular agency—in other words, there is a naturally strong case for the partnership. More likely than not, any effort to sole-source a project would need to be initiated by the agency, but agencies are not inclined to pursue this type of partnership regularly. There is a significant amount of rigorous internal-to-government justification necessary for such a partnership to withstand scrutiny.

In general, there is a significant amount of scrutiny *whenever* government funds are awarded in support of a project. That is true even when the funding amount is small. When I worked for the Department of Commerce, we partnered with a nonprofit organization to hold a conference—a seemingly simple activity—and we were investing only $25,000 in the event. A staff person in the Secretary's office raised questions as to whether our nonprofit partner was an appropriate partner because the organization as a whole took some positions that were contrary to the policy of the Administration at the time. To facilitate our sponsoring the conference, I had to go through three rounds of legal review to prove that the conference should be, in fact, a Department of Commerce activity.

Eventually, the conference was approved by the lawyers and was a successful event. Then, a couple of years later, the Department of Commerce was investigated *because of that same conference.* Fortunately, we had a thick file showing all the approvals received. I share that story to make the point that government entities have to take great care about which external entities they align themselves with no matter the activity, especially when it comes to ensuring that the projects they are funding align or are otherwise working in concert with the Administration's policy and that project activities are lawful and appropriate.

As a general principle, when taxpayer funds are being spent or when financial incentives or policies are being put into place on taxpayers' behalf, there are processes and practices to ensure fundamental fairness when it comes to how parties interested in partnering with government are treated. That also explains why most of the processes that help determine partnerships are competitive.

To maximize your chances of winning Federal funding, during your initial research phase, I recommend putting together a calendar identifying the specific programs and funding opportunities that you feel may be the best fit for your needs, along with a timeline and due dates. This programmatic review should be revisited once you have done your research, formulated your strategy, engaged with departments and programs that seem like a good fit for your work, and analyzed and prioritized from among those the ones best suited for partnership. Then you prepare your grant applications in an organized way, recognizing that it can take one to two hundred person-hours or longer to write a formal proposal.

Most people do not do this.

When organizations have approached TechVision21 for last-minute help, I have (reluctantly) turned the work down where I don't think they can be successful. In one case, we did decide to take on a last-minute project, but the process was very painful for all involved. Luckily, after a major effort, we got the job done, and amazingly, the client won an award. Know that their experience was an exception to the rule. A key lesson is to build in enough time for producing a solid proposal. The happy result for this client is not frequently shared by people who wait until the last minute to start work.

SHAPING FUNDING OPPORTUNITIES

In what follows, I review several considerations for approaching formal funding announcements from government agencies. But first, I want to review two opportunities for continuing to pursue a relationship with a government organization that can go a long way toward helping you shape agency ideas about what it should solicit in its formal requests for project proposals.

REQUEST FOR INFORMATION (RFI)

A Federal department or agency RFI is a solicitation of comments on its plans and goals. It presents an opportunity for people working in the field or on projects related to the field to join the conversation as well as offer points of view, information, and insights that the agency might have overlooked, failed to consider, or not considered thoroughly enough. Questions such as, What kinds of projects should we be funding? or What kinds of entities should be eligible to participate in this program? might be accompanied by questions about the general arena into which their work fits, such as, What would have the greatest impact on the industry as a whole? You may find that you are interested in responding to all or most of the questions posed, or that you want to respond to the one or two questions that are most relevant to your particular situation. Both approaches are welcome. Agencies read through all the comments they receive once the public comment period has officially ended. Sometimes, but not always, they publish a summary of the information they receive.

Whatever the agency leaders learned or considered as a result of the comments received frequently appears in different documents later on in the process. But it is a fair presumption that any agency that solicited information has taken into account at least some of the comments made by interested parties. Agency officials may not necessarily learn anything new, but they may discover gaps in their understanding or approach or come across novel research and justifications for considering shifting, enlarging, or shrinking their focus.

My organization worked with a large company to craft its response to an RFI. The Federal agency with which the company hoped to partner had left out of its list of questions any explicit consideration of approaches to energy efficiency of greatest interest to the company. In addition to responding to the RFI, the company's team briefed agency officials—in both cases suggesting that the agency might want to include this vital pathway to energy efficiency moving forward. Sure enough, when the agency's funding announcement was published, it included space for work on these issues. That may not always happen for every respondent to an RFI as easily as it did for this particular company, but the point is that agencies will make changes to what they are doing when they are convinced that it is appropriate.

Responding carefully to an RFI is one way for your organization to influence funding opportunities so that when a particular call for proposals is posted, it may be more of a fit for the work in which you are engaged.

BROAD AGENCY ANNOUNCEMENTS (BAA)

The Broad Agency Announcement is used widely by defense agencies and some others. The BAA is a way for the agency to categorize its interests. In many cases, the BAA can be seen as an invitation to visit and talk with agency representatives. In fact, the announcement may say as much outright, explicitly letting readers know that before they take any other actions, they should engage in a conversation during which they run their ideas by the organization's representatives. The BAA is also frequently a direct invitation to submit a concept paper. Depending on the agency and the funding cycle, representatives may even offer feedback. Occasionally, agency feedback can offer helpful hints about points of interest or focus that were not clarified within the BAA itself. The agency may or may not offer formal or informal feedback on discussions and written submissions.

It is certainly not guaranteed that an agency will respond to all the white papers it receives, but meeting with the agency in response to the BAA is always commendable. A given BAA usually has a broad window of influence, sometimes three, sometimes as many as five years. And if things should change for the agency during the period it has identified as covered by an existing BAA, it may replace its current version with another that better reflects any new direction.

On occasion, the BAA also serves as an invitation to submit a formal proposal and will provide explicit details on how to do that.

In other words, the BAA may lay out several stages all within a single announcement, with corresponding instructions for how to complete each stage. However, there may also be a separate funding announcement in addition to the BAA.

One of the best reasons to engage with an agency that has issued a BAA is that some agencies may have a predetermined sense of what they are looking for by the time they make public a funding announcement. As a result, there may be a short notice period during which the announcement appears to the public, making it difficult for those who are not already prepared to fulfill the proposal requirements in a short turnaround time.

For example, a program manager at an agency might send the public funding announcement to you by email, directly inviting you to respond, and then publish it more widely for a very short window of time. Different agencies are bound by different rules, and some may have short periods in which their funding announcements are available simply in order to avoid receiving a flood of proposals. Like the entrepreneur or organization looking for an agency that is a good fit for its ideas, so too are agencies, by way of their outreach methods, trying to increase their chances of getting proposals of the sort they most want.

FUNDING ANNOUNCEMENTS

Funding announcements go by different names depending on their source; they may be called Requests for Proposal, or as the Department of Energy calls them, Funding Opportunity Announcements (FOAs). They also have different processes and procedures for responding to them. As a general rule, you only want to respond *after you have tested the waters*—by meeting and talking with program managers or other decision-makers, or by submitting white papers or other communications. I always counsel clients that, although not impossible, it is difficult to win with a strategy of chasing published funding announcements when you have not previously met agency representatives and socialized your ideas. In a worst-case scenario, program managers may already have in mind the likely winners before they issue the public announcement.

Procedures vary from agency to agency, but in general, there are two major phases to the formal funding opportunity process.

1. **Concept Paper.** As I mentioned in our review of the BAA, an agency may invite the submission of concept papers as a first step in the proposal process. The request for concept papers might yield hundreds of entries, and a given agency might then invite only half or a third of those to submit a full proposal. That's a good thing, insofar as the early elimination round streamlines the process for both agency and proposers. First, it decreases the burden on applicants. Creating a five-page concept paper is itself hard work, as it involves clarifying your project definition, its impacts and nationwide benefits, your partners, if any, and your budget. But writing the concept paper is much less work than putting together a full proposal (often between fifty and a hundred pages or more) only to be rejected by the agency outright. Second, the two-phase process in which concept papers precede the full proposal helps the funding entity assess the extent to which an applicant or organization is "funding ready" or a good fit for the program. If the concept paper is unclear, its scope inappropriate, or its proposed project insufficiently justified, the agency has only to read through five pages, rather than a hundred, to make that

determination. Finally, if producing a good concept paper seems too difficult a task for your organization, that could be a sign that you are not ready to submit funding proposals in the first place.

2. **Full Proposal.** When it comes to requirements for the full proposal, the specifics of those requirements will vary by department or agency. Requirements may also vary in RFPs for the same department or agency, depending on the nature of the funding opportunity and anticipated grant size. That said, there will always be certain recurring elements across the board:

   A. Project Narrative
   The project narrative situates your project within the larger context of work in your area, offers a justification for undertaking the work, and thoroughly covers what you intend to do, how and where you intend to do it (and with whom, and using what equipment), and what the expected results will be if the project is successful. The expectation is usually that you already have access to the resources that you need, but sometimes some of what you need can be covered by a grant. In that instance, you will also be required to make a convincing justification for what you would like the government to provide, based on what you know about the resources that it regularly makes available to its funded partners.

   B. Technical Plan and Milestones
   Typically, projects will last between two and five years. It is imperative to be able to target within the proposal

what achievements along the way will help determine whether the project is on track and likely to be successful. The technical plan and identification of milestones may require the use of a Gantt or other chart tracking specific phases of the project. Some RFPs detail technical milestones that proposers are expected to achieve; in this case, you will want to explain in detail how your project will meet those milestones.

C. Budget

Your budget must be both realistic and reasonable. You should consider carefully how project milestones are tied to specific budget numbers so that you can expect funds received to match up with required expenses at appropriate times during the process.

D. Qualifications of Key Personnel

Not only must you acknowledge the people and positions that are necessary for completion of the project, but you must also make a convincing case that the people you have chosen are extremely well qualified to carry out the project that you propose.

E. Public Interest and Impacts

It is still important—perhaps even more important— at this phase to reiterate the significance of your proposed project, whatever qualifies it as novel or as an advance over the current state of the art and practice, and the potential impacts of the work both for the field and for the national interest. If you are proposing a project to the Department of Energy, they are likely to want to see that a successful project will yield impacts such as cost savings and increased energy efficiency. For example, given the billions of internal combustion engines used all over the world, if you have an improvement to the internal combustion engine that dramatically increases fuel efficiency and is cheaper to manufacture than what companies in the United States are doing today and accelerates time to development by 25 percent, your project is likely to have a dramatic impact, and you would not want to overlook, or underemphasize, this opportunity to say what your project could accomplish. Some agencies, such as the National Science Foundation, require proposers to identify "other impacts" beyond those stated in the review criteria. If there happen to be additional societal benefits to the project you are proposing, there is now available to you an opportunity to highlight them. For example, if the instructions about impacts have not offered the opportunity to name the environmental benefits of your proposed project, this would be the appropriate space in which to discuss those. Funding announcements—and their corresponding proposal instructions—typically include specific review criteria. It is very important to write to the agency's criteria in all sections of the proposal, especially regarding the sorts of impacts it is seeking. Do not ignore the stated criteria, even if the agency seems to be asking for something surprising, like regional geographic diversity among key personnel. Most important, do not attempt to fudge your ability to meet the stated criteria. Instead, figure out a way to meet them directly.

F. Corresponding Forms

These are often standardized forms that apply across a broad spectrum of funding agencies and opportunities. These include forms for budgets, subcontractors, and standard certifications (such as the acknowledgment that you will not pay a lobbyist using project funds). The important point here is to work the standard forms in parallel with the other parts of the application and not leave the forms until late in the process to complete. Have them drafted, reviewed (by your legal team or by senior project staff, as appropriate), and ready to go sooner rather than later.

G. Expressions of Support

On the matter of letters of support, there are typically two kinds: those from supporters (e.g., from the governor's office, economic development officials, or your Congressional delegation) who are eager for you to pursue your project but will not have a direct hand in it; and those from potential partners promising to make a specific commitment to your work (e.g., an organization that agrees to make a 10 percent in-kind donation of equipment and/or to perform a portion of the work). Seek these early in your proposal preparation process.

PREPARE OUTLINES AND ASSIGN TASKS

To ensure that none of the proposal requirements are overlooked or forgotten, I recommend beginning by combing through the BAA, the funding announcement, and all application instructions to create an outline for each separate section of the proposal itself. It can be helpful during drafting to have these explicit directives listed at the top of each section for everyone who touches the document to reference. Assigning responsibilities within sections to appropriate team members also helps ensure that everyone knows their roles: "Sam is in charge of forms. Sheryl is in charge of technical merit."

Then everyone who is helping to complete the proposal should meet regularly to review the sections, identifying where the holes are until there eventually are none left. Once drafts of the sections are all completed, the next critically important part of the work entails fitting all the different contributors' contributions together into a coherent and polished whole—a task, at least to start, for maybe one or two people with the skill set to ensure that everything hangs together and appears in logical order. Obviously, these master writers will continue to consult with, and seek comments from, the rest of the team. It is especially important during this part of the process to ensure that the proposal is written in a way that is understandable to nonexperts as well as to technical experts both within and separate from your specific field.

A final check will involve the following:

1. Reviewing the document as a whole with great care to confirm that every question asked has been answered and every instruction given has been followed as directly and completely as possible.

2. Ensuring that the document follows the order provided by the instructions.

3. Proposals that are incomplete or do not follow the rules may get tossed out of consideration at the very start of the review

process. A good proposal makes good sense fast, consistently proceeds in the suggested order, and addresses all matters requested thoroughly and completely and without reliance on jargon. As a general rule, getting a first draft completed as quickly as possible puts applicants in a great position to spend enough time on the usually complicated work of refining and improving the proposal.

Putting together a proposal requires a considerable amount of work—I would describe it as a month's worth of effort for at least one person, if not two or three. That difference will depend on the complexity of the agency's proposal submission process and the number of parties involved. I cannot stress enough the importance of allotting enough time for recruiting partners and others whose help you will need to actually perform the work. Especially in larger companies, running permissions up the chain can be very time-consuming. Maybe you have contacted someone on the technical staff at a given company who is happy to participate in your project, but then that person needs to get the official go-ahead from five separate layers of management in order officially to participate. Allot time for just these sorts of delays.

A final word on proposal submission, one which I know to be both obvious yet very little heeded: do not wait until the last day or the last minutes before the deadline to submit your proposal through the agency's system. I came close to having a client miss a deadline once because they had too many people on the team responsible for different proposal components and did not start to put them together until a half hour before the deadline. There were about six people sitting in a room together during that final window for submission, and they called me on the phone, frantically trying to file the different components. I found myself advising on assembling the proposal in real time over the phone without being able to see the proposal components. Then there was no way to check the submission after the fact, so I spent months afterward worrying that they would be disqualified due to an incomplete proposal. Fortunately, and against all odds, the proposal was complete. And the client won an award. Do not follow their lead! Give yourself, at the very least, in excess of twenty-four hours to submit your proposal, and prepare a checklist that you can use to help you through the process. Websites crash, proposal pieces might be discovered missing, and as-yet-unnoticed errors may appear.

If you have not taken advantage of opportunities to present your ideas before and there is an agency solicitation of any kind, you will have to do your best to prepare a proposal in good faith, ensuring that you check all the boxes when it comes to meeting requirements.

When writing a proposal, I recommend that you plan ahead. It should be standard practice, in most cases, to spend time and energy writing a proposal *only* when you have done the homework of meeting people, learning goals, and assessing how competitive you are. Remember: By the time the funding announcement comes out, you are competing with people who have been paying close attention and who have engaged in careful planning and meeting with program managers.

TAKE ACTION

1. Make it an explicit part of your strategy to respond directly to Requests for Information and Broad Agency Announcements as a means of helping to shape funding opportunities.

2. Begin the proposal-writing process by creating an outline that follows the order of sections as they are expressed in the funding announcement and that clarifies precisely what content needs to appear in each section.

3. Give your team enough time at all phases of the process, including time to acquire and confirm the participation of key personnel, time to complete all the corresponding forms, and time to ensure that you complete the submission process well before the official deadline.

Politics is more difficult than physics.

—ALBERT EINSTEIN

CHAPTER 7

PARTNER TO INFLUENCE POLICY

ONE OF OUR CLIENTS had a nanotechnology-based product that was a natural fit for use by the military; however, the first thing this company's leaders learned when they approached the government with their product was how difficult it is to get on the Federal government's approval list to sell to the military. There is quite a process for getting your product on the military purchasing schedule, and part of that process requires testing and demonstration that what you are offering in fact does what you have said it will do.

The reason for the testing and demonstration should be clear enough: the United States military is not going to be open to using an untested, unverified, and largely as-yet-unutilized product when it already has perfectly reliable products that have proven useful for the past twenty or more years.

The company went through several introductory meetings to explain its problem and eventually got to a point where there was a good bit of interest—so much interest, in fact, that the company was able to secure a meeting with a dozen people in the government who might be prospective purchasers. The company representatives presented their product to the prospective purchasers and talked about its performance and results. They showed off their data. The conversation seemed to be going well.

Then, because it is imperative that the government have very rigorous processes for approving new things, the group from the government asked for more data on specific use cases. This particular company had the data for which it was being asked, but its leadership was uncomfortable sharing data. This particular company could have become approved as a government supplier if it had recognized that the government representatives simply needed to ensure the legitimacy of every organization's claims. And government partners generally have an obligation to protect data and information they receive from prospective partners.

Just as you have to be willing to address biases rather than take offense at them, so too, you have to be willing to meet the demand for further proof if you intend to establish a functional relationship.

I have already mentioned the need to anticipate the biases into which you might run when you present something innovative or revolutionary to a Federal agency. That said, there are sometimes occasions when the government's own safeguards can prevent new products from finding a market. I mentioned earlier working with a team that was developing a new internal combustion engine around the same time that the DOE wanted to promote electric vehicles as a more economical and environmentally friendly way of meeting its goals. In this instance, my clients eventually succeeded in making the case that there was more than one way to meet the department's goals and that the nonelectric engine they were developing could achieve those goals and then some. In the end, they helped start a conversation around achieving fuel efficiency and environmental goals in a more cost-effective way by looking at technologies outside the scope of those currently tagged by DOE as most desirable.

To this point in our review, I have focused on pathbreaking, innovative technologies. Sometimes what is innovative about a technology may in fact be clear enough, such as, when it contributes dramatically to improvements in a well-determined and well-known product or process. But when new technologies do something different from what has come before—or solve a problem previously believed to be impossible—or are based on ideas imported from another field—people naturally may not understand them. In these instances, some people may be skeptical that what is being proposed can be accomplished. Beyond those individual resistances, government regulations and laws tend to be written in support of the status quo, sometimes preventing truly new projects and products from gaining appropriate consideration.

That is one of the reasons why I have placed so much emphasis on the importance of understanding the current environment for your technology, product, or service. Knowing precisely how what you do responds or relates to what is already being done is a necessary part of making the best case for your work. In some cases, especially if you hope to influence policy, it may become necessary to conduct a broad-ranging educational strategy as part of your outreach efforts to government agencies. You may first need to set the scene and encourage the government to consider the benefit of entertaining the technology options you have to offer.

For example, during the period when the lithium battery was preferred by the Obama Administration, one of our clients made a high-performing lead acid battery. As part of their plan to seek government partnerships, the client first needed to persuade agencies to consider battery options other than those using lithium. In other words, they had to make the case that they should be allowed to compete for funding and partnerships even though their product was not aligned with those most sought after by the then-current Administration. Our team worked with them to develop a compelling message, backed by data, and they briefed widely and well. Over the course of about eighteen months, they were able to make a lot of inroads with critical decision-makers and identify pathways to commercialization.

It certainly does not help matters that laws and regulations frequently operate in the same silos as does R&D. For example, the solar and wind industries have their own sets of financial incentives and tax credits and programs to support development and deployment. Nor does it help that policymakers and legislative officials have their own working assumptions about how each technology performs and which ones are superior. Personal assessments, coupled with access to particular studies or sources of information, can end up influencing what is written into law. That means both that you may have a skeptical audience for your product or service and that you may be able to shift points of view and share information that changes the way the law works.

As part of my own efforts to influence the way government processes function when it comes to technological innovation, for a long time now, I have been making the case for technology-agnostic decision-making across departments and agencies. If the goal of a particular agency or program is to reduce CO_2 emissions in the atmosphere, I have argued that instead of favoring one method or one type of product over another, the best way to achieve this is to include in funding opportunities and other partnership opportunities, a general requirement for a specific amount of carbon reduction, and/or a particular level of efficiency and performance, and then let competitors figure out how to best meet those standards.

A good illustration of the opposite—what we might call technology-committed decision-making—sometimes occurs in the arena of environmental regulations. President Nixon's creation of the EPA was the first time the government placed environmental regulations and requirements on industry. These regularly took the form of commandments—"You should do this; you should not do that"—and were organized by specific environmental media: air, water, soil, and so on. The introduction of these measures had a significant impact: everyone raced to meet the new standards, and performance improved across the board.

But then, over time, once regulated entities were regularly meeting those standards, the standards became the floor. That raised the following problem: unless the government keeps raising the floor and increasing regulatory pressures on industry, people do not have as much incentive to perform any better. That well describes how things proceeded for quite some time, but now that the EPA is again taking a regulatory approach and creating higher standards, the scene is shifting once again. California already has been a leader in raising standards at the state level, and several other states followed their lead.[29] Now, California's goals are beginning to be adopted at the Federal level. The push toward net-zero emissions vehicles by the year 2035, for example, will drive the market toward battery-operated vehicles and crowd out other kinds of propulsion—that might be just as efficient, or even more efficient—simply because those other kinds of vehicles are not favored in the regulations developed under certain legislation.

---

29   In April 2023, the Biden Administration proposed new Corporate Average Fuel Economy Standards that require an industry-wide fleet average of approximately 49 miles per gallon for passenger cars and light trucks beginning in model year 2026. The Administration also announced its intention to tighten emissions rules on both new and existing power plants.

I count not just R&D funding opportunities but also demonstration opportunities within the scope of decisions that should be made in a technology-agnostic way. Some government departments and agencies, such as the Department of Defense and the military services, run their own demonstration programs. Traditionally, if an agency decides it needs to demonstrate a particular technology, such as solar, because demonstration is frequently a necessary step toward moving the technology into general use, the agency needed to bring that idea before Congress and make the case for why Congress should appropriate the funds for the demonstrations to proceed. The appropriations cycle itself can be long, often taking between a year and two years and, in this case, only affects solar technologies.

To me, that is a very inefficient process. The policy would be greatly improved by having a demonstration program in which all technologies can participate. Once again, departments and agencies would identify the results they hope to achieve, allow any kind of project to compete, and then have program managers select the best projects from across the portfolio of all technologies in which the department or agency invests.

The good news is that the Biden Administration adopted a technology-agnostic approach and created a demonstration office at the DOE within the Office of Energy Efficiency and Renewable Energy to fund projects across the various technology silos within the portfolio of government interests. This is a thrilling development, but it does not eliminate what I take to be an ongoing problem that organizations and entrepreneurs face: they may not be able to obtain government support because there is not yet a welcoming environment for that particular technology.

That said, the policy environment can be changed or at the very least influenced. Efforts to create change might take place at the agency level, at an Administration-wide level, or likely, both. There are even Administration-wide bodies—such as the Office of Science and Technology Policy and the National Science and Technology Council—that have a coordinating role across Federal R&D funding agencies and make interagency reports and recommendations.[30]

It is easy enough to point out that change can occur, but changing the policy environment is not easily accomplished. Doing so involves convincing a lot of people, as there is no one person or decision-making entity that can determine a new policy. In theory, the President of the United States could do that so, but even an Executive Order does not guarantee that all the people working within and across agencies will quickly change their ways. Some may change quickly; others may never change at all. Most importantly, Congress often has the final say.

That is why there is great value in briefing up and down the hierarchy and across partner agencies when it comes to influencing the policy environment. If you want a program to run differently, you talk with the people who run the program—typically those people are career senior executives—but you also talk with the political, policy, and programmatic officials at the relevant agency. Then, too, you work with people on Capitol Hill, because they control the purse strings. You may also wish to brief outside of government to influence organizations that help to educate decision-makers and advocate for changes in policy.

My team has worked with several clients seeking to provide analysis, products, or services beneficial to military service members. In every case, engagement with outside organizations representing service members proved decisive in securing funding and government alliances.

---

30   There also are many other formal and less formal coordinating bodies that work within and across agencies. For example, when I was in government, we had a very active "deputies" council which worked to resolve issues before escalating them to the Cabinet Secretary level. These types of bodies are still important features of the Federal landscape today.

If you decide to take this approach, there is value in looking at a few years' worth of authorization legislation and reports to see what direction a potential partner agency has gone in and where it is going, and then assess whether that direction potentially helps or hinders your attempts to partner with that agency. It also is imperative to review at the relevant appropriations legislation to see exactly how much money has been and is being directed to which programs and projects.

PARALLEL PROCESSES

In earlier chapters, we reviewed the value of doing one's research on specific agencies and programs to track their stated objectives. That early research goal was aimed at learning enough about an agency or program so that you can comfortably begin briefing people about your project and assessing their interest. Here, in our review of the policy environment, our research reference point is shifted slightly. When we are talking about influencing the policy environment, the group of people with whom you schedule meetings is likely to shift—you may be dealing with the same agencies but seeking conversations with different people within those agencies, or you may be talking with people beyond those who influence decision-making at the agency level. In addition, when you attempt to influence policy, you may make different points than when you meet with representatives to argue the value and significance of a particular project.

These processes run in tandem and parallel to one another. At the same time that you pursue a specific relationship, you might also submit written comments that go into the public record to make the case for how the agency as a whole should approach the subject area in which you work. You might talk with policy people about what you consider to be the best kinds of projects they could pursue within your area of expertise. Influencing policy requires a shift in perspective from the specific to the more programmatic, a consideration of the broader arena into which any of your projects might fit.

Sometimes the Requests for Information that a department or agency releases will themselves read as very policy-oriented. A recent RFI from the Department of Commerce, reviewed the five responsibilities the Department had been given under a current statute, then posed a set of questions asking for recommendations on how it should fulfill those five responsibilities. Besides general questions on best approaches and timelines, there were questions about more specific areas of concern. For example, Congress directed the department to create a National Science and Technology Center to focus on prototyping. So Commerce asked corresponding questions about the best kind of consortium to create for the center and ways to facilitate the center's work with other programs. Similarly, there was a section on workforce development, asking readers to identify workforce needs, kinds of development programs that would be most beneficial, and ways of embedding workforce development within other areas of the department's efforts.

POLICY AWARENESS

In most cases, partnering with government will not require you to attempt to change the policy environment. Instead, what will be most important is that you are keenly aware of what is occurring at the policy level so you can be prepared to take advantage of time-sensitive opportunities as soon as they arise. Some Federal programs are formulated as block grants to states, leaving individual states to determine their priorities and desired outcomes in conjunction with meeting the Federal objective. One current program in this arena is called the National Electric Vehicle Initiative, and one of its primary concerns is to develop a charging infrastructure to support the use of electric vehicles. For more than a decade, there have been one-off projects around the country that have largely focused on fleets of buses and service vehicles that drive all day and then return to a depot to charge at night. But recently, attention has been given to addressing the concerns most people have about buying electric vehicles. Besides the expense, many potential buyers have "range anxiety" and are concerned about whether they will be able to use electric vehicles for longer-distance, highway drives. "If I'm driving from Washington, DC, to Detroit, Michigan," they ask, "where will I charge my vehicle along the way?"

Current Federal policy in this area, originally proposed by the Obama Administration, instructs the Departments of Transportation and Energy to collaborate to create a national charging infrastructure. Under the Biden Administration, the Federal government established a program offering block grants to states and mandating charging stations every fifty miles within a corridor around interstate highways. The individual states are responsible for implementing that policy in whatever ways make the most sense for them. As I write this, all fifty-six states and territories are preparing to solicit proposals that meet both the Federal set of standards along with each individual state's own related priorities and standards. Any entity that wishes to compete in more than one state will need to get quickly up to speed on what the policy issues are at both the Federal level and state levels.

Another aspect of policy awareness involves understanding which initiatives and goals the current Administration is talking about, even bragging about. To return to the example of the CHIPS Act, President Biden has regularly talked about how the act is helping to move the country forward, promoting innovation, and providing funding to ensure that the United States is a leader in semiconductor technology. He drew attention to the ubiquity of semiconductors—their status as the foundation of our TVs, our cell phones, our washing machines, our cars. Paying careful attention to the issues that are repeatedly emphasized and the rationales being put forward in support of initiatives to address those issues is one way of positioning yourself as a match for current and future efforts.

I recommend that companies and entrepreneurs balance their efforts between the pursuit of specific projects, maintaining ongoing and up-to-date policy awareness, and pursuing efforts aimed at policy change or expansion where needed.

In all, I recommend that companies and entrepreneurs balance their efforts between the pursuit of specific projects, maintaining ongoing and up-to-date policy awareness, and pursuing efforts aimed at policy change or expansion where needed. Sometimes you will find yourself focused on making your project better fit the directives and goals of a given agency or program; at other times you may need to work on changing the policy environment in which those goals and directives take shape. There may even be instances where both efforts could be combined; half your presentation might focus on making the case that your project is a good fit for a given program and the other half might focus on offering suggestions within the larger policy environment. What your messaging entails depends, of course, on the people with whom you're in conversation.

Ultimately, your message gets noticed and your work assisted when you're in conversation with the people who can help make that happen. We have talked about identifying the appropriate deci-sion-makers, but I would remind us here that sometimes there are people with quite a bit of influence on those decision-makers who should not be left out of your briefing and relationship-building strategy. When it comes to policy, as with specific projects, your goals should be the same: refine your message, be thorough in your search for the appropriate people with whom to talk, and make the rounds to agency and Congressional representatives and to other parties with interests in these issues, such as think tanks and industry associations.

BRIEFING UP, DOWN, AND ACROSS

The most frequent policymakers are the political appointees at each of the departments and agencies. They are also the people who tend to be most in tune with what each Administration has identified as its priorities. But these appointees are likely also to check in with and take seriously the insights provided by the people who run the programs themselves and who have perhaps been running those programs for quite some time. Depending on the sort of change that is being considered, it may be the case that that change can be made at the agency level, or it may be the case that the agency lacks the legal authority to make the change and needs to talk with Congress before anything can be done. The key is to understand as much as you can about the process and find all the champions you can within a given agency, within Congress, and with external players, such as think tanks and industry associations.

For example, if you are working in technologies, you may at some point see value in talking with the chairperson and ranking member of the Congressional Energy and Commerce Committee. You might look at the members of the subcommittee that deals with solar power to see if any of your state's representatives are on it. You might talk with the committee staff who work under the direction of the chair, and so on.

There is not necessarily any right order for approaching people with different roles and at different levels of government authority—with one caveat. I have seen a lot of people start by visiting their Congressional representatives' offices and asking for recommendations regarding whom they should approach at a particular department or agency. I do not usually find that a successful way to proceed and suspect that many agencies respond less well to being approached by people leaning on the authority of a Congressional representative to get their foot in the door. It has been my contention in this book that if you hope to partner with an agency and to compete for funding from that agency, you start with the agency and do all the things that I mentioned earlier to introduce your work and build relationships there. As an agency representative, I strongly believed it was an important part of my job to meet with people and hear about their work, so I view this step as relatively easy and straightforward. Then, you may decide to visit your Congressional representatives to let them know what you are doing and, if suitable, ask for letters of support that can be included in your funding proposals.

What is most important to a productive partnership with government is to view the government as your ally and the relationships you are seeking and establishing as having long-term and lasting benefits. Like any relationship, there is work involved in maintaining and strengthening your government partnerships. Like the best relationships, there are benefits to be shared between the partners over the course of a lifetime.

I told a story at the start of this chapter about a deal that fell apart in which both the company and the nation lost out on a great opportunity. That result can also occur when would-be partners of government think solely in terms of their organization's business needs and do not also take seriously the matter of addressing national needs. In point of fact, one must take both into consideration. To the question, How can this help me meet my business needs? you need to add, How can this help my nation?

TAKE ACTION

1. Consider the extent to which it may be necessary to conduct a broad-ranging educational strategy as part of your outreach efforts to government agencies and Congress.

2. Identify and consider when it might be most appropriate to talk with people across government agencies and up and down existing hierarchies who have influence over the policy environment.

3. Stay abreast of the policy environment, including reviewing Administration priorities as well as budget authorization and appropriation legislation.

4. Consider whether working with like-minded organizations outside of government—such as think tanks and industry associations—can help amplify your message and expand your reach.

It is the long history of humankind (and animal kind too) those who learned to collaborate and improvise most effectively have prevailed.

—CHARLES DARWIN

CHAPTER 8

BUILD LONG-TERM PARTNERSHIPS

A CLIENT WITH WHOM I've worked in several capacities over the years and helped establish partnerships with the Federal government had the idea to create a facility that would bring state-of-the-art semiconductor fabrication capability to the State of Florida. At the time, he was working for a nonprofit organization that had successfully obtained a commitment from the County for the project. But to make the organization's and County's vision a reality, a significant investment was needed. When we started working together on this project, the three things he had to show off were some photos of undeveloped land in Central Florida, an artist's rendering of the anticipated $200+ million final facility, and about seven slides detailing how the facility he imagined would fill a unique niche within the industry's prototyping and low volume production processes.

We spent at least a year meeting with people in Washington, from the Departments of Energy, Commerce, and Defense to relevant Congressional committees, laboratories, think tanks, other nonprofits, and just about anyone else who might be interested in the project. After over a year of meetings, the idea started getting traction, and he was able to convincingly demonstrate the viability of the building he envisioned. Over time, the organization recruited a visionary CEO, who tirelessly promoted the vision. Thanks to a substantial investment by the State of Florida and some Federal interest, five years later, the undeveloped land had been replaced by a magnificent building that would not have come to be without both the tactical strategy we employed and the perseverance the clients brought to the business of convincing others—at all levels and in all places—that they could bring to life something that, initially, almost no one thought was possible.

My overall argument to this point has been focused on demonstrating how to go about making the government your long-term partner. Once you have proved your bona fides to them, established your relationships, and perhaps even won a couple of funding competitions, you should strongly consider ways to continue building on those relationships. In the case of the clients I mention here, the track record that we had helped them establish helped position them as a trustworthy partner in whom additional, and larger, investments might be made.

As an established partner, you can approach the government with new ideas, joint opportunities, and potential collaborations to bring products to market.

As an established partner, you can approach the government with new ideas, joint opportunities, and potential collaborations to bring products to market. You can also lean on these established relationships to enlarge your community of collaborators. There are a number of defense agencies, for example, the prime contractors of which have mentor-mentee programs with small businesses. If you're a small business with which a government agency has worked successfully, your agency partner might connect you with a mentor who is a prime contractor in their program.

Similarly, given that agencies almost always have their own sets of contractors with whom they work regularly, those contractor organizations and their people can become champions for your work. Agencies with which you partner over the long term are also likely to let you know about conferences they are hosting or in which they are participating, or they might ask to use your project as a case study or sample to promote their programs and sponsorships. Given that the advantages can be quite numerous, I think that every company, every organization should strive to reach this level of long-term relationship with government.

I've so far mentioned here benefits besides those that are monetary, but longer-term relationships are also likely to yield higher levels of monetary investment from government agencies. Once you have become a trusted partner, your work with government can grow over time into larger and larger contracts. It can even grow into its own line of business.

In an earlier chapter, I mentioned the Small Business Innovation Research Program, which is limited to companies with fewer than five hundred employees. I have worked with small businesses that have grown their partnerships with government agencies over time—from Phase I funding, which is a relatively small amount of money, to Phase II (a larger amount of money)—then worked with an agency like the National Science Foundation, which has a Phase IIB opportunity, offering a little extra money to help prepare for commercialization. When these small businesses were ready for

Phase III funding, the commercialization and sales phase, they were placed into a program of record with a larger agency for them to provide their product as one component in a much larger system. What is key to establishing that sort of connection is that programs of record frequently receive line-item funding from Congress; so once these small companies were participating in a program of record, they had a reliable means of supporting their work and providing it with steady clients. Some of them would never have gotten to the point of commercialization—let alone gotten their businesses off the ground—without that long-term Federal support.

To be clear: if you do a good job with the smaller opportunities that you are offered, and if you are transparent and open about what you are doing by sharing your test data and treating the government department or agency as a true partner, you build the foundation upon which that government entity will remain eager to grow its relationship with you. Earlier, I talked about growing your relationship with the Federal government as akin to establishing and nurturing a friendship. That is even more salient a recommendation at the point where you have in fact formed a working relationship and perhaps even multiple specific friendships within a given agency. Those relationships help you establish a platform to build on for future success—whether that's further R&D or work on product commercialization—and they need continuous nurturing.

In the prior chapter, we reviewed the benefits of partnership with government when it comes to influencing the policy environment in favor of your projects and plans. That same point holds here in terms of long-term benefits. Once government departments or agencies know that you are a trustworthy partner, once they believe that you have a good solution to problems they are facing now or anticipate facing in the future, they may shape their policy agenda so that it creates an environment generally beneficial to your technology.

Once you win over the confidence of the very people you need to help make your case, you can sustain a relationship from which, quite naturally, changes to the policy and regulatory environment are likely to follow.

It is worth considering that one of the most productive benefits of sustaining a long-term partnership is that it offers yet another avenue for overcoming some of the inherent biases within the system. Earlier, we acknowledged how difficult it can be when you have a new product to offer, something pathbreaking, only to discover that the very people who might help you bring it to the world hesitate to trust it. We noted that one path by which to address this obstacle is to argue your point to the right people at the right time. Another path is to establish a solid working relationship. Once you win over the confidence of the very people you need to help make your case, you can sustain a relationship from which, quite naturally, changes to the policy and regulatory environment can follow.

This is even more the case when your policy goals affect more than one department or agency. Officials from one agency may be able to persuade their colleagues at another to make room for your technology in their work. Along these same lines, your agency contacts may be able to help you both on Capitol Hill and with other business opportunities outside of government, including nonprofit organizations and potential business and university partners who have similar interests and could benefit from establishing a relationship with you.

In the end, long-term relationships with government can be mutually beneficial and open up a whole new set of relationships, even a whole new community with which you interact. Even if you already have been working toward establishing that community as part of your business plan, close partnerships with government can solidify your position as well as expand your network to include people you may not yet have discovered on your own or not yet imagined how your work might overlap with theirs.

There is still one other way of pursuing long-term partnerships with the Federal government, and it occurs when organizations are able to recognize when the government is attempting to achieve internally what is more easily achieved, or has already been achieved, in the private sector. Ideally, we would have a robust information marketplace in which businesses can better understand Federal needs and agencies can complement and support one another, rather than duplicate work that's already been done. But given that we are still working to achieve such a marketplace of ideas and information, it is still the case that there are occasions when the government may be unaware of the most efficient means of reaching its goals.

For all the government's efforts to ensure that it is enabling technological advancement, there are sometimes better tools already in existence that have been developed privately than could be developed publicly. Recognizing that you have a better-performing product than what the government has been working to produce internally is both valuable to government and an opening to a fruitful partnership. By introducing concepts and developing alliances with government agencies, you can not only help create a positive environment for the work you are doing and for your products and services, but you can advance national goals by bringing the government better solutions than it has been able to find on its own.

To the extent that we all work, at least in part, in our own worlds, in our own silos of activity, so too can the Federal government. It is just as important, if not more, to broaden the government's view as it is to inform the general public about your ideas and offerings. And unlike those cases in which you may be proposing an idea or solution that does not fit whatever paradigm the government is currently working with, in this case, you have a very practical offer to make: *we already have that capability, and we are ready to sell or share with you a version of it that can help meet the nation's immediate needs.*

I acknowledge that for those just considering government partnership, or for those whose attempts at partnership have to date been met with less than the warmest reception, it can seem like working with government amounts to a series of lessons in delayed gratification. Frankly, the major reason people stop attempting to work with government is that they become discouraged that they are not successful as quickly as they would have hoped. I understand how that experience can occur, and I recognize that by comparison with other avenues, the government can offer a longer-term cycle to get a product developed. And unless you are extremely lucky, it is difficult to circumvent the time cycles according to which government processes operate.

Nevertheless, I have written this book to help convince those working on the cutting edge of a wide gamut of technology fields that there are substantial rewards to be reaped by building and sustaining a trusting relationship with government partners. Not the least among these rewards are the possibility of eventually being connected to large funding streams and the longer-term likelihood of being introduced to a new and ever-expanding community of potential collaborators. The "seal of approval" you can earn by working successfully with government can help open up and accelerate other private market opportunities. Most important, some projects just may not be attractive to private investors for a variety of reasons.

The payoffs of government partnerships can surpass what organizations and entrepreneurs can achieve by seeking private-sector funding.

As a final word on the matter, I offer this: the payoffs of government partnerships can surpass what organizations and entrepreneurs can achieve by seeking private-sector funding. Venture capitalists are no longer in the habit of throwing big dollars at every idea out there, and VC funding can be fickle if anticipated results don't immediately materialize. In the current economy in particular, and in other moments when the economy has leveled out a bit, venture money is less readily accessible. Furthermore, there are phases in the technology R&D process that are not necessarily amenable to venture funding.

Put otherwise, venture capital frequently cannot give you everything you need to bring your product to market.[31] I have witnessed companies arrive at the pilot testing phase of their new technology only to discover that they are not in a position to return to the VC market. They may have received a few rounds of funding initially, but they cannot return to the same funding sources to raise enough money to move forward. Supplementing an organization's progress with governmental relationships can be a wise choice in those moments, both in terms of funding and in terms of overall support and promotion of ideas.

I encourage you to add the government to the list of serious potential partners you consider as you work toward making the best business decisions at various points in the product development process. This may mean that you find it useful to seek partnerships with government in the early phases of your research, or perhaps you will find it beneficial to partner with government further along in the

---

31   According to the *Wall Street Journal*, as I write, other than investments in AI, "most of the Silicon Valley's venture-capital ecosystem remains in the doldrums..." AI Startups become magnets for money, *Wall Street Journal*, May 9, 2023

product development and deployment process. Some of the best relationships I have seen have been ones in which companies transition in and out of more focused government partnerships depending on what suits them well at a given time. That is still another good reason for keeping one's relationships with the Federal government alive and intact over the long term.

As someone who has worked for government, worked in the business world, and now established businesses, universities, and nonprofits find creative ways of forming government partnerships, I am convinced that those who put their minds to it and strategize carefully and thoroughly about how to maximize partnership potential can work very successfully with government in ways that both shape the environment in which they are trying to operate and help build a platform on which they can generate future success.

TAKE ACTION

1. Be transparent and open with your government partner(s) to maintain good standing, prove your trustworthiness, and move your products and services forward.

2. Decide how government partnerships best fit into your long-term business strategy, identifying the ideal times to call on your government partners for assistance.

3. Be proactive about growing and maintaining your relationships. Remember that the government truly wants to help new technologies reach the public.

THE AMERICAN INNOVATOR

IN MY COMPANY'S work helping innovators establish and sustain productive relationships with the Federal government, I try never to lose sight of two important details: the government wants to support projects that serve the national good—whether promoting security, health, or a clean environment; ensuring a stable and safe food supply; promoting high-quality job opportunities; or addressing climate change—and many innovators are driven by a desire to contribute to the national or global good.

In the corporate world, there is an incentivization process referred to as "doing well by doing good," which encourages the production of products and services that ultimately provide for the common good. That motivation pairs well with the mission of the Federal government. From the long-term partnerships between government and industry that I have both participated in and witnessed, doing good work that makes an impact is the truest satisfaction drawn by both partners.

I acknowledge the reluctance felt by some regarding partnering with the United States government or with government contractors. A recent conversation with my son brought just this problem to mind. He and I both want our work to be meaningful, as we assume is true for many people earning a living in America today. He is convinced that batteries are world-changing applications of technology in which the United States may be lagging behind other countries and expressed to me how much it would mean to him to help develop and commercialize next-generation energy storage technology in America. But my son also shared the idea that his continued engagement in just this sort of meaningful work might necessitate avoiding organizations that are long-standing government partners and contractors.

Of course, I was eager to disagree. More precisely, I was eager to note that "big government contractor" organizations are composed of multiple divisions and with many capabilities. He could work for a large organization within a division that focused only on battery technology. Doing so would ensure that he contributed no effort to the technologies being produced by that organization to which he was averse or with which he disagreed. Moreover, given the dangerous world in which we live today, with a ground war in Europe and potential military threats from Russia and China, among others, I believe deeply that supporting America's national defense is a noble and necessary effort.

Where he decides to work is, of course, his own business, but I take seriously the point of view that he expressed along with some people's distrust of the establishment. But it would be unfortunate, in my opinion, if that sense of distrust led our brightest minds to miss out on opportunities to pursue meaningful work and make a measurable difference in people's lives by partnering with a government aimed at doing just the same. If there is one message I hope you take from this book, it is that there is deep potential value in finding ways to pursue partnerships with the United States government.

The clients I most enjoy working with are ones that see how establishing a relationship with the Federal government can catalyze their efforts to do well by doing good. I appreciate, too, when clients approach partnership with the Federal government as an ongoing journey worthy of their continued effort. With these clients and those who feel similarly, I emphasize the importance of finding early on people to integrate into your team who can take you all the way through that journey, from initial research and strategizing to plotting effective communications across a variety of audiences to finding ways of further enhancing successful, established partnerships.

Incorporating onto your team people who bring experience working with multiple agencies and multiple processes is one of the very best ways to position your organization to see across and think strategically about the entire government portfolio and policy environment. Being able to look broadly and deeply at available partnership opportunities in the short, medium, and longer term—not just across multiple government agencies but also including private-sector partners—is ultimately the best way to ensure that you establish a plan that maximizes available opportunities for your company.

Not the least of reasons the United States is unique on the world stage is that our founders crafted a Constitution that explicitly fosters experimentation, invention, and technology development.

As at the start of this book, at the end I remind you that America is a country built on innovation. Not the least of reasons the United States is unique on the world stage is that our founders crafted a constitution that explicitly fosters experimentation, invention, and technology development. In the history of this nation, technological innovations have catalyzed tremendous economic growth, job creation, industrial competitiveness, and entrepreneurial success. Still today, our nation's long-term economic growth is deeply intertwined with its support for cutting-edge technological advances.

The Federal government has a long-standing commitment to funding innovative technologies in areas of benefit to the nation, even, and especially, when it comes to funding research that is considered too expensive or too risky to be supported by private investment. The involvement of the Federal government in accelerating development and commercialization of the most innovative technologies is undergoing a major boost right now, one that will set the stage for as-yet-unimagined innovations for decades to come. Now is one of the best times in the history of the United States to ask and begin answering this question:

What role will your company and your projects play in the continued success of our nation?

ACKNOWLEDGMENTS

I AM GRATEFUL to the numerous friends and colleagues with whom I have worked closely over many years on many types of projects, including Anita Balachandra, Carol Ann Meares, John Sargent, Katie Wolf, Michael Daum, and Carolyn Van Damme, all of whom have taught me the value of research, creative thinking, and successful collaboration.

I have fond memories of my time working for the Federal government and of the professional colleagues who so generously shared their wisdom and advice with me over the years. I am thankful for the influence of our leaders at the Department of Commerce Technology Admininstration, Dr. Mary Good, Dr. Graham Mitchell, and Dr. Arati Prabhakar; Commerce Secretaries Bill Daley and the late Ron Brown, Deputy Secretaries Dave Barram and Robert Mallet; my colleagues Cheryl Mendonsa, Virginia Miller, Kent Hughes, Judy Jablow, Gary Bachula, Cathy Campbell, the late Jon Paugh, Dr. Phyllis Yoshida, and many others.

My gratitude also extends to the clients with whom we have worked at TechVision21 and those with whom I've had the pleasure of developing long-term collaborations. My admiration extends to all the people working in and outside government to advance innovation and American competitiveness.

Nathan, Jason, and Haley Carnes, your support and encouragement over the years have been invaluable gifts. Without them, I might never have considered—let alone made time for—producing these pages.

Printed in the USA
CPSIA information can be obtained
at www.ICGtesting.com
JSHW011252171023
50337JS00004B/106